HISTORICAL GEOLOGY OF INDIA

S.K. Shah
Former Head,
Department of Geology
University of Jammu
Jammu - 180 015

SCIENTIFIC PUBLISHERS

Published by
SCIENTIFIC PUBLISHERS (INDIA)

Jodhpur –

5 A, New Pali Road
P.O. Box 91
Jodhpur - 342 001 INDIA

Delhi –

- 4806/24, Ansari Road
Daryaganj
New Delhi - 110 002 INDIA

ISBN: 978-93-86652-12-6

Visit the Scientific Publishers (India) website at
http://www.scientificpub.com

Printed in India

CONTENTS

PREFACE

History of the earth is a long saga dating back to more than four and a half billion years. The earlier part of its story is still misty and clouded with assumptions which are yet to be fully understood and explained. However, the later parts, notably the last billion years, are better documented, and we have fairly good idea of the happenings. This is mainly because the record of the remains of organisms is better preserved and throws a light on various aspects of the evolution of the earth's lithosphere and biosphere. The historical evolution is revealed through stratigraphic studies.

Stratigraphy is a rapidly expanding branch of Earth Sciences wherein the application of new and precise techniques is changing interpretations. Consequently, there is a greater potential of certain accepted principles being revised from time to time, and some of them becoming redundant. For the students of Earth Sciences, it is a challenge to remain abreast with the changing concepts and apply them for regional studies. The challenge becomes more difficult in the absence of text books, wherein the application of new principles to regional studies is included and the latest developments are recorded.

The students of Earth Sciences in India have been handicapped for need of up-to-date text books in stratigraphy and historical geology of India. For most of the 20th century, the main text books available were those by D. N. Wadia and M. S. Krishnan. The former became outdated by the 60s, while the latter was obsolete by the end of 70s. Since then, no serious effort has been made to bring out a text book which could replace them. The attempts made are either in the nature of general books on Indian geology which have little bearing to the curriculum of graduate and post-graduate studies or outdated because of some spurious data that found its way in them following a fossil scandal. Recently, the Geological Society of India brought out a book on the Geology of India in two volumes. This is an appreciable effort by the authors. These volumes could be of use for researchers. However, they cannot be used as text books as per the curriculum of various universities in India.

During my experience as a teacher of stratigraphy and regional geology of India spanning for nearly four decades, I have always found myself running a losing race. While the information of studies conducted in different regions of India was increasing, the same could not keep pace with the rapidly expanding concepts of stratigraphy. I realised alarmingly, that I did not know how to teach the historical evolution of India from a stratigraphic perspective. One of the main handicaps was the absence of state-of-the-art text books. The students invariably resorted to outdated text books for writing their answers. They failed to relate classroom lectures to what was available in the text books and, therefore, preferred to depend on the latter. Some of the more resourceful ones who tried to consult other literature such as published papers also failed to separate grain from the chaff and would put in some subjective information of individual authors. Irrespective of what was given to them in class lectures and seminars, it was invariably the same story when they came to writing answers.

During my interactions with other teachers spread over the country, I found that all of them faced the same problem. In fact, it was reflected on national-level competitive examinations such as those of the Union Public Service Commission and other examinations conducted by government, semi-government and non-government organizations. The need to bring out a text book that would resolve this problem was immense.

Since stratigraphy is a dynamic science, it is necessary to be clear about the basic principles. While the enunciation of these principles is outside the scope of the present volume, a summarised version of the same has been included as the first chapter so that the students may not need to search for other sources to understand the broad applications.

One of the major problems related to regional stratigraphic studies is the explosion of formational names that has taken place during last few decades. While there is a code of nomenclature in operation, the subjectivity of these names is enormous, even where all nomenclatural rules have been followed. This is mainly because the principles of correlation are rarely followed. While theoretically all the names that are proposed as per code of nomenclature are valid, their use places a burden on students who have to put to memory an unnecessary plethora of oddly spelt and bizarre sounding place names. No wonder the subject is tending to become boring if too much emphasis is put on using these names. Objectively for a student it is not important to remember the sequential details but understand the implications on local, regional and global level in a time concept. Accordingly nomenclatural details have been included only where they are necessary.

While talking of formational names there is another problem related to it. Over the decades certain names have got deeply entrenched in literature and they are well understood. However, since code of nomenclature was proposed much later, some of these do not fit into it. Accordingly many attempts have been made to replace them without much success. A teacher is faced with a dilemma as to whether he should use the old name or the new one. Since it has not been possible to popularize the new name nor is there uniformity in its application, it is preferable to use the old name in an informal manner since it is well understood. In this volume the same practice has been followed.

The bibliography appended is not exhaustive and is only meant for students and research scholars to refer to some important source material in case they need the details on a particular subject. Generally those publications are listed that are of a review type or contain some original material that is relevant to the subject.

The standard stratigraphic scale incuded as Appendix I is based on Geological Time Scale as approved by the Commission of Stratigraphy of the International Union of Geological Sciences. It only covers the Phanerozoic part of the column since that is better understood and systematically correlated. The details of the Precambrian sequence are still sketchy and they are included in the main body of Chapter I.

The writing of this book has been a modest effort by somebody who claims to be a teacher but finds himself still a student. In spite of best efforts some inadequacies are bound to be there in this volume. All the teachers and students are requested to point out the same so that they can be rectified in future. Any other suggestions for the improvement of the book are welcome.

S.K. Shah

Jammu
August, 2017

1 Stratigraphic Concepts

1.1 STRATIGRAPHIC PRINCIPLES

Stratigraphy is that branch of Earth Sciences which deals with the sequence of stratified rocks in relation to their time of deposition. The stratified rocks are the repository of a record of past events. Stratigraphy deals with sediments and sedimentary rocks, sedimentary environments and fossils that have a direct bearing to time of deposition. The fundamental criterion of stratigraphic study involves the principle of superposition. According to this principle, the older layers or beds in a stratified rock sequence are below, followed upwards sequentially by younger beds. Moreover, superposition includes the principle of original horizontality and original continuity. The former means that although the rock strata may often be found in an inclined condition, they were originally deposited in a horizontal position. The latter means that although the rocks may not occur as continuous outcrops, they were originally continuously deposited in a basin. The inclination is a result of later deformation, whereas the discontinuity of the strata in the valleys or elsewhere follows the processes of erosion. The relative time of deposition of each bed can be determined accordingly through the order of superposition. This helps in classifying the rock sequence on the basis of time of deposition. Stratigraphic study aims to classify and correlate rock sequences to the time of deposition in order to understand the historical development of the Earth and reconstruct its evolutionary history.

All stratigraphic studies have a space and time component. The space component involves geographical distances or lateral distribution. The vertical sequence is directly related to time. Therefore, the classification of time is an important component of stratigraphy. The relative time is determined and correlation undertaken on the basis of fossils in case of rocks deposited during the last 542 million years, where fossils are abundant and well preserved. It is this last part of the Earth's history that is referred to as Phanerozoic, literally meaning the time span bearing visible life. For older rocks and others, where fossils are poorly represented or absent, other non-palaeontological techniques are applied. These non-palaeontological techniques include superposition, petrographic characters, breaks in deposition known as unconformities, structure and tectonics. Furthermore, various geophysical techniques such as electrical logging, palaeomagnetism and seismicity are widely used for classification and determining the stratigraphic sequence and correlation.

Since rocks are exposed in different geographic and structural settings, it is necessary to correlate them to obtain a continuous stratigraphic order. Therefore, correlation constitutes an important theme of stratigraphy. The purpose of correlation is to match rock formations of different areas with respect to the various stages in the Earth's evolutionary history. These stages may or may not be synchronous (deposited at the same time) but they contain identical fossils or have a similar stratigraphic setting. When identical fossils occur in widely separated areas, it does not necessarily mean that the deposition has taken place at the same time. There may be a time lag in dispersal of organisms from one place to another. Thus, identical fossils in geographically distant areas can also be dichronous (deposited at two different periods of time).

Stratigraphy deals with layered rocks deposited in a sequence. This sequence is indicative of time of deposition. Therefore, there are two fundamental components to stratigraphy: rock sequence and the relative time that the sequence indicates.

The rock sequence follows the principle of superposition; this is the reflection of time during which this deposition has taken place. The relative time as interpreted from the rock sequence is reflected in the Geologic Time Scale that has been proposed for the objective measurement and study.

1.2 UNIFORMITARIANISM AND CATASTROPHISM

In the 19th century, when the Earth's history was not well understood and stratigraphy was still in a state of infancy, a great debate was taking place as to whether the changes on the Earth were a result of major catastrophes due to sudden upheavals or gradual evolution due to the normal processes operating on the Earth's surface. The former school of thought was referred to as catastrophism and the latter as uniformitarianism. While there were supporters for both schools, uniformitarianism became a more accepted mechanism by the end of that century. The principle of uniformitarianism was that the processes as currently operating on the Earth such as sedimentation and erosion were the same in the past; moreover, these processes over a prolonged period of time were responsible for all the changes that have taken place on the Earth. The guiding maxim for understanding the historical evolution of the Earth under this principle was that 'study of the present is the key to the past'.

As the science of stratigraphy evolved and a better appraisal of the processes that have been operating in the past became available, this debate has become redundant. Most of the geological processes operate very slowly, at least in terms of human life span; therefore, uniformitarianism can be considered instrumental in bringing about changes during the prolonged periods of the Earth's history. However, catastrophic events have taken place even in human history, and there are innumerable evidences of such events in the geological past that have completely transformed not only the tectonic disposition of the Earth but also the nature of living beings. Most mass extinctions of organisms on the Earth are a result of such catastrophes. Hence, the Earth's history consists of long periods of gradual and uniform changes followed by occasional short periods of a catastrophic event that disturbs a relatively quiet and slow evolution. The demarcation of such events has helped us in classifying the stratigraphic sequences into major divisions since these events are based on a time control. Consequently, larger divisions of the Earth's time span are based on faunal and floral changes due to catastrophic events, whereas most minor divisions are a result of the changes in organisms through uniformitarian processes of evolution. Thus, while both uniformitarianism and catastrophism are the operating processes in the Earth's history, the major changes are all a result of the latter.

1.3 GEOLOGIC TIME SCALE

The stratigraphic sequence of layered rocks is classified by a scale that has evolved through the correlation of rocks in different settings. This geologic time scale is primarily a relative time scale and is not based on the age in absolute years. The age in absolute years indicated against each scale is based on radiometric dating that has subsequently come up and is still in the process of revision. The last 542 million years—called Phanerozoic time—is better classified since it bears fossils and can, therefore, be easily correlated. The older rocks referred to as Precambrian constituting the Proterozoic and Archean are not so well classified since their correlation is based on non-palaeontological techniques.

In the Geologic Time Scale, the first order unit of rock sequence is called a **System**. A System is a basic time-rock unit that denotes the rocks deposited in a particular period, as indicated by the fossil content. The corresponding time unit during which the deposition of these rocks took place is called a **Period**. A System is further divided into **Series** for which a corresponding time unit is an **Epoch**. Given below are the time-rock divisions in descending order and their corresponding time divisions:

Time-rock units	Erathem	System	Series	Stage
Time units	Era	Period	Epoch	Age

The time-rock and time units in the scale are denoted by the same terms, but the suffix indicates if it denotes time-rock unit or time. For example, the Cambrian System denotes a time-rock division while the Cambrian Period denotes time division during which those rocks were deposited. Tables 1.1 and 1.2 show the Phanerozoic and Precambrian Geologic Time Scales, respectively.

1.4 STRATIGRAPHIC CLASSIFICATION

Stratigraphically, rocks are classified in different ways depending on the nature of the sequence and the purpose for which the classification is undertaken. Two types of classifications—namely, lithostratigraphic and biostratigraphic—are based on observed data, the former on the nature of the rock types and the latter on the fossil content. A third kind of classification—namely, chronostratigraphic or time-stratigraphic—is based on the interpretation derived from the biostratigraphic or other data and constitutes the basis of the above-mentioned Geologic Time Scale.

Table 1.1 PHANEROZOIC TIME SCALE

(Note: The divisions of Cambrian are informal and are yet to be ratified by International Commission on Stratigraphy)

ERA/ ERATHEM	PERIOD/SYSTEM (SUBPERIOD)		EPOCH/SERIES	AGE(Ma) (at the beginning)
CENOZOIC	QUATERNARY		HOLOCENE	0.011
			PLEISTOCENE	2.58
	NEOGENE		PLIOCENE	5.33
			MIOCENE	23.03
	PALEOGENE		OLIGOCENE	33.9
			EOCENE	55.8
			PALEOCENE	65.5
MESOZOIC	CRETACEOUS		LATE	99.6
			EARLY	145.5
	JURASSIC		LATE	161.2
			MIDDLE	175.6
			EARLY	199.6
	TRIASSIC		LATE	228.7
			MIDDLE	245.9
			EARLY	251.0
PALEOZOIC	PERMIAN		LOPINGIAN	260.4
			GUADALUPIAN	270.6
			CISURALIAN	299.0
	CARBONIFEROUS	(PENNSYL-VANIAN)	LATE	307.2
			MIDDLE	311.7
			EARLY	318.1
		(MISSISSI-PPIAN)	LATE	328.3
			MIDDLE	345.3
			EARLY	359.2
	DEVONIAN		LATE	385.3
			MIDDLE	397.5
			EARLY	416.0
	SILURIAN		PRIDOLI	418.7
			LUDLOW	422.9

		WENLOCK	428.2
		LLANDOVERY	443.7
	ORDOVICIAN	LATE	460.9
		MIDDLE	471.8
		EARLY	488.3
	CAMBRIAN	FURONGIAN	499.0
		SERIES 3	510.0
		SERIES 2	521.0
		TERRENEUVIAN	542.0

Table 1.2 PRECAMBRIAN TIME SCALE

EON	**ERA**	**PERIOD**		**AGE (Ma) (at the beginning)**
PROTEROZOIC	NEOPROTERO-ZOIC	EDIACARAN (=VENDIAN)		635
		CRYOGENIAN	(=STURTIAN)	850
		TONIAN		1000
	MESOPROTERO-ZOIC	STENIAN	(=RIPHEAN)	1200
		ECTASIAN		1400
		CALYMMIAN		1600
	PALAEOPRO-TEROZOIC	STATHERIAN		1800
		OROSIRIAN		2050
		RHYACIAN		2300
		SIDERIAN		2500
ARCHEAN	NEOARCHAEN			2800
	MESOARCHAEN			3200
	PALAEO-ARCHAEN			3600
	EOARCHAEN			4000
HADEAN (informal)				4600

1.4.1 Lithostratigraphic classification

Lithostratigraphic or rock-stratigraphic classification is primarily based on the rock types that constitute a sedimentary sequence. The basic unit of this classification is a **Formation**. A Formation consists of a homogeneous set of strata comprising a distinct rock type or set of rock types which is laterally continuous and of sufficient thickness so that it becomes a mappable unit. The homogeneity may be by way of lithological or other distinctive characters such as colour or weathering characteristics or a combination of these. Each Formation may be further classified into **Members**. Likewise, several Formations may be bunched together into a **Group** and several Groups into a **Super-group**. Thus, various units in lithostratigraphic classification in descending order of thickness are Super-group, Group, Formation and Member. However, the primary and mandatory unit of this classification is Formation alone.

The lithostratigraphic units are named on the basis of locality or geographical features such as a river or a mountain peak, where a type section is exposed. A type section or **stratotype** should include the full complement of the formation. The locality or geographical feature name is suffixed by unit name. The examples would be the units such as Barakar Formation and Cuddapah Supergroup. In case of Formations, the name of the rock type that constitutes the unit may also be used as the suffix. Examples of this would be Muth Quartzite and Rohtas Limestone. A code of stratigraphic nomenclature is in existence that must be strictly followed in assigning names, and each new formational name should include the full description of the stratotype and a stratigraphic column.

1.4.2 Biostratigraphic Classification

The basic principle of biostratigraphy is that the evolutionary changes in organisms are non-reversible; therefore, the fossils present during a time span are characteristic only of that particular time interval and cannot repeat or replicate at any other time. Accordingly, on the presence of characteristic fossils, the strata are divided into **biozones** which constitute the basic unit of biostratigraphic classification. There are various types of biozones depending on the nature and distribution of fauna and flora. The most commonly used biozones are **Range Zone** (also called **Acrozone**), **Assemblage Zone** (also called **Cenozone**) and **Abundance Zone** (also called **Acme Zone**).

A Range Zone is based on the range of a fossil taxon, i.e. the span between its first appearance and final disappearance. For fixing a Range Zone, only those fossils can be utilised which have a short time range and a maximum possible widespread geographic distribution. Such fossils are known as **index fossils**. On the basis of the total ranges of such fossils, the biozones are fixed. In the absence of index fossils, the strata are classified into Assemblage Zones. An Assemblage Zone is defined as a biozone characterised by the association of many taxa (usually three or more). Only on the basis of the combined association of these taxa can the zone be identified and not on the presence of any single fossil taxon, as in the case of Range Zone. An Abundance Zone is defined as a body of strata that contains a maximum population of one or more species. In other words, the taxa involved were at their acme or climax; therefore, this biozone is also called Acme Zone. Since the acme of any species may be based only on local palaeo-ecological factors, Abundance Zones are used only for local classification and do not have a widespread connotation for correlation purposes.

For biostratigraphic classification, there are usually no stratotypes but reference localities can be named. A Range or Abundance Zone is named on the basis of the generic or trivial name of the index or acme species, respectively. In the case of an Assemblage Zone, the name can be based on the two of the assemblage taxa. The Zones may be further divided into **sub-zones** or further into **zonules**, the latter being the smallest biostratigraphic sub-division.

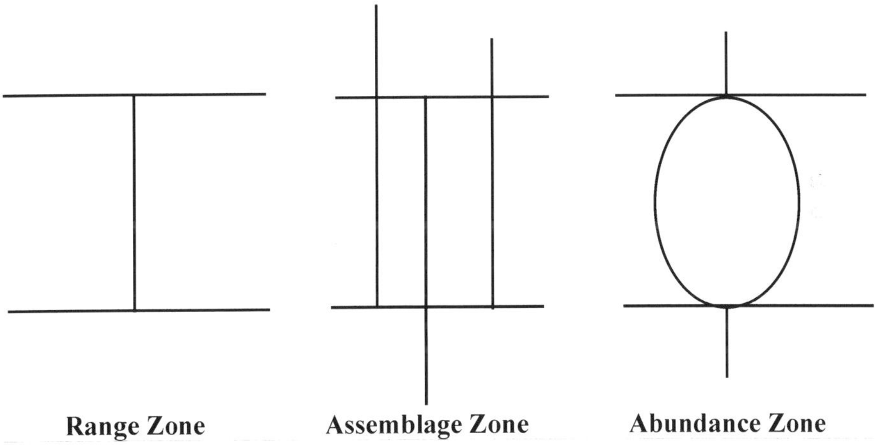

Fig. 1.1 Types of Biozones (Vertical lines denote the fossil taxa)

1.4.3 Chronostratigraphic Classification

Chronostratigraphic classification is an abstract concept and is based on time as interpreted from biostratigraphy to a large extent. Therefore, biostratigraphy is intimately connected to chronostratigraphy, and the latter is primarily dependent on the former. The Geological Time Scale is basically a chronostratigraphic scale. The time encompassed by a biozone is termed as a **biochron**. Moreover, on the basis of biochrons, the time unit called **age** and its corresponding time-rock unit **stage** (see Table 1.1) are defined. Chronostratigraphic units are the fundamental working units of the Geological Time scale.

1.4.4 Sequence Stratigraphy

In recent years, a new concept of detailed lithologic sequence that has a time connotation has been developed, which is referred to as sequence stratigraphy. In sequence stratigraphy, the rock deposits are subdivided into unconformity bound units. These unconformities may be based on the variation in sediment supply and changes in the relative sea level or other surfaces that may represent time lines due to various events such as seismicity, subaerial erosion or flooding. These unconformities are a result of events at a particular point of time. Therefore, sequence stratigraphy represents chronostratigraphic classification based on cyclicity of sedimentation which follows well-marked events in time.

Sequence stratigraphy is completely different from lithostratigraphy which does not have a time connotation. Lithostratigraphy is based purely on rock types; however, the same rock types may be deposited at different points of time or different rocks may be deposited at the same time, depending on the nature of the basin. Sequence stratigraphy is not based on rock types but on events displayed in the nature of sedimentation showing unconformities due to variations which occur at particular points of time. These sequence boundaries mark the different units in a sedimentary succession and these units that are genetically related can be correlated.

1.5. CONCEPT OF FACIES

The sediment types deposited in a basin are determined by various factors: notably, sediment supply (or what is known as provenance), nature of the basin (marine, lacustrine, fluviatile etc.), water depth, turbulence, fauna and flora, water chemistry and many other variables. Altogether, they constitute what is known as the environment of deposition. The term **facies** is applied to a particular environment of deposition. However, basically the term facies means aspect and simply refers to the character of the rock type. In a same basin, as we laterally trace a rock type, it may change its character. For example, a sandstone unit when laterally traced may become shaly and eventually pass on into shale. This would evidently imply a change of environment from shallow to deep water. Likewise, any other environmental change in deposition would bring about a corresponding change in lithology or fauna and flora or both. The term facies is used to denote a specific environment of deposition. It may be characterised by a particular rock type when it would be called as **lithofacies** or characteristic fauna and flora when it would be denoted as **biofacies**.

The facies concept has a bearing on the lithostratigraphic and biostratigraphic classification discussed above. The various methods of classification mentioned above, cannot be used interchangeably, though they have a short-term relationship restricted to limited geographical areas. The lithostratigraphic classification is purely based on lithology; however, within a limited area, each unit may represent deposition at the same time. However, as the facies changes, we may be confronted with two different rock types deposited at the same time in a basin. With the variation in facies, the same rock type may be deposited at different points of time. Furthermore, the same basin may deposit one rock type at one place and another at a different place as the change of facies is a common phenomenon in a basin. In other words, lithostratigraphic units invariably cut across time as the distance increases since the facies in any basin is not identical throughout. Most rock units are time-transgressive; therefore, it would be incorrect to treat them as time units. Likewise, difference of fossils in the same basin may not have any bearing on the time of deposition as they may be purely facies controlled.

1.6 CORRELATION

Correlation forms the basis of stratigraphic classification. The purpose of correlation is to determine the identical time-rock units on a local, regional and global scale to facilitate uniformity of stratigraphic classification. Therefore, correlation involves the matching of rock sequences as exposed in different areas. The scope of correlation can vary depending on its nature. On a local scale, correlation can be undertaken simply on the similarity of lithology and related structures. Accordingly, the lithostratigraphic units can be correlated from different areas within a limited geographic setting. However, this may not be possible on a regional scale because similarity of lithology is not necessarily indicative of deposition at the same time as it is more a reflection of the nature of the basin and other environmental factors in different geographical regions.

Regional and global correlation is primarily undertaken with the help of fossils, and it involves the correlation of the various kinds of biozones. While most biozones are of a local significance depending on the nature of the fossils, there are some fossils that have a widespread distribution and can thereby help in regional correlation. For example, fossil groups such as ammonites, graptolites, some planktonic foraminifers etc. are usually widely distributed and constitute some ideal index fossils for regional and global correlation. However, one caution needs to be always kept in mind: the similarity of fossil taxa, or what is known as homotaxis, on a regional scale does not necessarily mean contemporaneity or deposition at the same time. There is always a time lag of dispersal of organisms and greater the distance more is this time lag. There can also be physical and environmental barriers which may impede the dispersal. The dispersal potential of different organisms varies, determining their utility for intercontinental correlation.

Precise time correlation is possible through radiometric dating which is based on the principle of decay of radioactive elements, which takes place at a fixed rate. However, this would not give the time of deposition of a rock and cannot be used for most of the sedimentary rocks but can be indirectly used by determining the ages of any intrusives or volcanic rocks associated with the sediments. For extremely young sediments, radiocarbon dating is a highly precise and useful tool. Other methods which help in time correlation involve palaeomagnetic studies by determining the magnetic reversals in time and correlating them on a global scale.

1.7 RADIOMETRIC DATING

The radioactive elements in rocks decay and disintegrate into what are known as daughter elements. The rate of this disintegration is constant under all conditions. If the rate of disintegration is known, the proportion of the atoms of both parent and daughter elements present in the rock will determine the time elapsed in the formation of the mineral that contains the radioactive element concerned. In this manner, the absolute age of the mineral can be calculated. This constitutes the basic principle of radioactive dating.

The rate of disintegration of an isotope is quantified by what is known as the half-life period. Half-life period is the time taken for the decay of half the atoms of the parent element. Half-life periods of most isotopes are in thousands of million years and, therefore, they can be used for dating old crystalline and volcanic rocks. The most common isotopes that are used for dating such rocks are Potassium (K-40), Rubidium (Rb-87), Uranium (U-238, U-234 and U-235) and Thorium (Th-232) that disintegrate respectively into Argon, Strontium and Lead (both Uranium and Thorium).

For the dating of extremely young rocks the most commonly used isotope is Carbon (C-14) which has a half-life period of only 5730 ± 40 years. It is used for the dating of those rocks that contain organic carbon derived from atmospheric carbon dioxide. This dating can be used only up to a maximum of 40,000 years because the amount of C-14 remaining is minute and undetectable beyond that time. Radiocarbon dating is different from the other radioactive methods of dating in that it is related to the processes operating in atmosphere and hydrosphere and not with the crystallisation of minerals. In the upper atmosphere, the C-14 isotope is produced from nitrogen-14, which gets incorporated in the carbon dioxide and eventually finds its way in the oceans and living plants and animals. In nature, there is a balance between the two through recycling. The recycling of C-14 continues during the life of the organisms but ceases after its death and its decay commences. Radiocarbon dating depends on the measurement of remaining C-14 in the dead organic material.

The radioactive damage phenomena are also used in other techniques of dating minerals in young rocks by recording the trail of damage through the surrounding mineral matter. These trails can be studied under high magnification after suitably etching the rocks and are known as **fission tracks**. The degree of density of the tracks determines the age. However, the tracks get annealed and fade away if the rock is heated, thereby implying that fission track ages only give the date since cooling and not the true age of the mineral.

For dating of extremely young sediments and historical and archaeological remains, one of the commonly used modern techniques which are based primarily on radiation damage involves thermo-luminescence and electron spin resonance. These techniques involve the measurement of energy in metastable electrons that are displaced during radiation damage.

Radioactive dating is one of the fastest growing areas with the advancement in techniques and is a highly specialised field involving sophisticated instrumentation. While it is outside the scope of average stratigraphic studies, it has made a significant contribution in understanding the contemporaneity of events in Phanerozoic stratigraphy. For Precambrian stratigraphy, it constitutes the backbone for understanding the sequence of events.

1.8 MAGNETIC STRATIGRAPHY

Magnetic stratigraphy is a relatively recent technique that has opened a new and reliable means of time-correlation of sediments. It is based on the fact that the Earth's magnetic field reverses itself at intervals, and record of the original magnetic field at the time of the deposition of the sediments can be seen in them as what is known as residual magnetism. This magnetism is recorded in sediments bearing magnetic minerals such as those of iron and is better available in fine-grained sediments deposited under low energy conditions. It also gets recorded in lava flows. What happens during a reversal is that the magnetic field strength drops briefly to a zero value before increasing again with opposite polarity. Reversal has occurred many times in the geological past. For example, during the last five million years, four epochs of reversal, two with normal polarity and two with reversed polarity, have taken place. The residual magnetism can better be recorded in relatively younger rocks. The sequence of magnetic polarity records for the entire Phanerozoic sequence has now been built up, affording a highly reliable time-controlled record of global events that can be used for dating rocks and precise time correlation of sedimentary rocks and lava flows. The polarity data is shown as **polarity chrons** that comprise a set of normal and reversed time intervals. Earlier, the magnetic polarity time scale (MPTS) was shown in running numbers for each polarity chron. With the availability of greater details in recent years the pattern of numbering has been revised to provide a more precise division. The magnetic polarity data is more accurate in younger rocks. Accordingly, Cenozoic rocks have well-documented polarity chrons covering the whole of Cenozoic and a little part of Late Cretaceous. There are 34 sets in all—numbered as C1 to C34 from the top (C standing for Cenozoic)—with the normal and reversed sub-chrons suffixed with N and R, respectively. For Mesozoic rocks, the polarity chrons are in M series. However, for older rocks, the data is poor and the numbering of chrons has, therefore, not been formally attempted.

Another application of palaeomagnetism is in plotting the position of poles that vary in time in relation to any assumed fixed point of the Earth's crust. This variation is recorded as apparent polar wandering curves (APWC). It is seen that apparent position of magnetic poles has been continuously changing through geological time. APWC not only helps in locating the position of different blocks or plates of the Earth's crust at various points of time but can also be used as means of dividing geological time on a larger scale.

■■■■

2 Physiography of India

2.1 TECTONO-STRATIGRAPHIC DIVISIONS OF INDIAN SUBCONTINENT

The Indian sub-continent comprises two well-marked physiographic divisions with a distinct tectonic history, viz. the southern Peninsular region comprising the Indian Shield and the northern Extra-Peninsular region constituting the Himalaya and the adjoining alluvial tract. The former is one of the oldest shield areas of the globe, which was stabilised in Precambrian after a series of orogenic cycles, attended upon by phases of metamorphism and granitisation. As against that the Extra-Peninsular region is still in a state of mobility and underwent the orogeny only in Cenozoic times. Accordingly, the stratigraphic and tectonic history of the two regions is considerably different.

While appreciating this difference between the two tectonic regions, it must be emphasised that their historical evolution cannot be studied in isolation. The two regions constitute a single geographic entity and the history of the one has affected that of the other. Eventually, during the Himalayan orogeny, they got welded together and even the northern margin of the peninsula got involved in this orogeny. Accordingly, several rock units of Lesser Himalaya are the re-mobilised elements of peninsular India which constitute the basement for the deposition of frontal Himalayan Cenozoic sediments including the Indo-Gangetic alluvium. Therefore, the boundary between the two regions cannot be sharply drawn. Otherwise also, the Indo-Gangetic alluvium and the desert region mask a great part of the northern margin of the peninsular rocks. The nature of the basement can either be indirectly inferred or is known from some bore holes drilled for the location of oil and gas.

2.2 HIMALAYAN MOUNTAIN BELT

The name Himalaya is derived from the Sanskrit words *him-alaya* meaning snow-covered ranges. Since it is already pluralised, it is incorrect to call it 'Himalayas' as is the usual practice with western writers. The Himalayan range is one of the longest and highest mountain chains in the world. It extends for over 2500 kilometres from Pamir in the west to Mishmi Hills in the east. The mountain range has a width ranging from 150 to 400 kilometres. Some of the highest peaks of the world, including Mount Everest (8848 metres), K2 (8611metres) and Kanchenjunga (8598 metres), are located in this range. The mountain range is an arcuate chain with convexity to the south. The range passes through several countries extending from Pakistan in the west through India into Nepal, Bhutan and Tibet. On the western and eastern ends, there are two knee-bends known as syntaxial bends. The western syntaxial bend is around mount Nanga Parbat while the eastern bend it is at Namcha Barwa.

The Himalaya can be divided linearly into four divisions depending on altitude and other topographical features. The divisions from south to north constitute Outer Himalaya (also known as the Siwalik range), Lesser Himalaya, Higher Himalaya (also sometimes referred to as Greater Himalaya or the Sanskrit name Himadri) and Tethyan Himalaya (sometimes, though erroneously, called Tibetan Himalaya). Outer Himalaya comprises low hills that gradually rise from the Indo-Gangetic Plain and attain an average height of 1000 metres and thereabouts. It has a width ranging from 10 to 50 kilometres. It is followed by Lesser Himalaya that has a height range of 1000 to 2500 metres. This region is highly rugged and tectonically disturbed. Its width is approximately 50 kilometres. The Higher Himalaya that succeeds Lesser Himalaya has a height range above 2500 metres with extremely high peaks going even beyond 8000 metres. The average height of the peaks ranges between 6000 and 8000 metres. This region constitutes the watershed for the consequent drainage of all the Himalayan river systems in the Indian sub-continent except for Indus, Sutlej and Brahamputra which cut across it. Higher parts of this region are perpetually snowcovered and all Himalayan glaciers are located in this part. It is the widest part of Himalaya with width ranging from 120 to 190 kilometres. Beyond the Higher Himalaya is the trans-Himalayan region which is known as the Tethyan Himalaya in the absence of any suitable topographical nomenclature. It consists of a plateau region over 3000 metres in height with a gentle slope northwards which continues into the Tibetan plateau.

The different sectors of Himalaya from the western syntaxial bend to the eastern syntaxis are named as Kashmir Himalaya, Himachal Himalaya, Kumaun–Garhwal Himalaya, Nepal Himalaya, Bhutan Himalaya, Arunachal Himalaya and Assam Himalaya. Topographically, there is not much difference between these sectors and the nomenclature is only for purposes of description.

2.3 PENINSULAR INDIA

2.3.1 Plateau region

As compared to the Himalayan region, the main peninsular part of India comprises low lying plateau area with stumpy mountains projecting out. The northern part of the peninsular region is covered by Indo-Gangetic alluvium constituting what is known as Northern Plain that separates the peninsular area from the Himalayan region. The peninsular plateau is enclosed within the low mountain ranges. To the north-west, it is bounded by Aravalli chain and by the Vindhyan Hills and Satpura chain to the north. On either side of the triangular part of the plateau are the Western and Eastern Ghats.

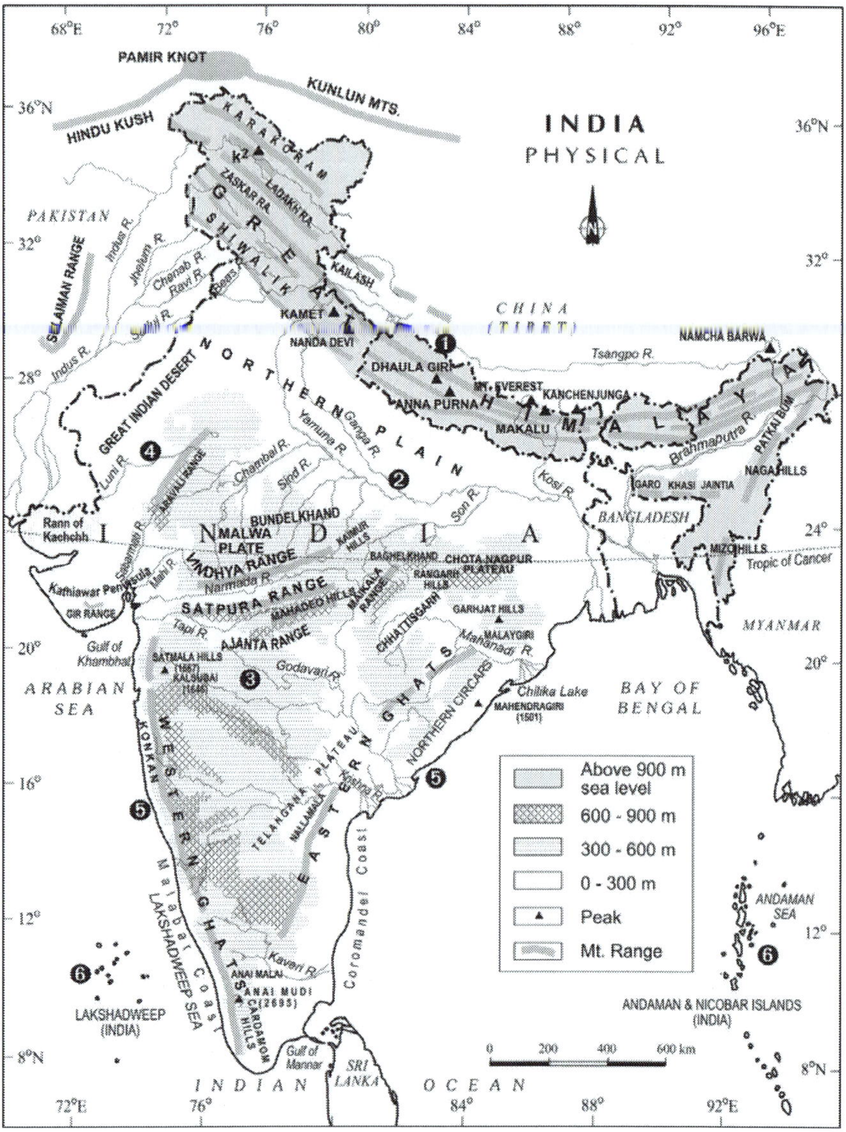

Fig. 2.1 Physiography of India: 1. Himalaya. 2. Indo-Gangetic Plain (Northern Plain). 3. Peninsular Plateau. 4. Thar Desert. 5. Western and Eastern Ghats and outer coastal areas. 6. The Islands. (Adapted from the website of UPSC General Studies and Current Affairs https://iasmania.com/Physiographic-Divisions-of-India/)

The Peninsular plateau can be divided into two parts: namely, Central Highlands and Deccan Plateau. The two are separated by the river Narmada that marks a major tensional tectonic boundary. The northern Central Highlands bear the Aravalli mountain range that extends from Gujarat through Rajasthan to Delhi. Comparatively, it is a high mountain chain in peninsular India and the highest peak in the Mt. Abu region (Gurushikhar) is 1722 metres high. The other regions of Central Highlands constitute the Malwa and Chhotanagpur plateau. As mentioned earlier, the triangular Deccan Plateau is bounded by the Western Ghats, Eastern Ghats, Vindhyan hills and Satpura chain of mountains to the north. The Western Ghats (also known as Sahyadri chain) extends from South Gujarat to Kerala and comprises low hills with an average height less than 1000 m. The highest peak (Anaimudi) is approximately 2693 metres high. The Western Ghats is a continuous chain that constitutes the watershed of the rivers Godavari, Bhima, Krishna and Kaveri. As against the Western Ghats, the Eastern Ghats comprise a series of detached flat topped hills with an average altitude of 600 metres. The highest peak (Mahendragiri) is 1501 metres high. The range extends from Orissa to Tamil Nadu with major breaks in between. The Vindhyan chain to the north comprises low mountains and plateaus that extend from Gujarat in the west to Bihar in the east and has an average elevation of 500–600 metres. The Satpura range is short in length but higher that Vindhyan range with the highest point (Pachmarhi) attaining a height of 1335 metres.

2.3.2 Coastal Plains

The coastal plains in India run along the Arabian Sea and Bay of Bengal. Along the Arabian Sea, the coastal region lies to the west of Western Ghats. It is relatively a narrow belt (10 to 20 kilometres wide) stretching from Rann of Kutch to Kanyakumari. The different sectors of the coast are variously named on the basis of geographical location like Konkan Coast, Karnataka Coast and Malabar Coast. The northern-most part of the coast constituting the Rann of Kutch gets partly submerged under the sea during high tide and becomes an estuary during Monsoon period but emerges as land in winter months. The entire coastal plain is a low-lying area. It is a site of deposition of sediments derived from the Western Ghats. However, there are also erosional landforms due to wave action.

The eastern coastal plain along the Bay of Bengal is much wider than the western plain. It has an average width of approximately 120 kilometres. It is marked by a series of deltas of rivers flowing into the Bay of Bengal: namely Mahanadi, Godavari, Krishna and Kaveri. Moreover, there are a considerable number of lagoons in between the deltaic regions. The northern part of the coast is called Northern Circar and the southern part is called Coromandal Coast. Eastern Orissa and Bengal coastal region constituting the Northern Circar is especially full of lagoons in addition to the large deltaic area of the Gangetic system of rivers. There is also the Sundarbans estuary which gets submerged during the monsoon season like the Rann of Kutch.

2.4 INDO-GANGETIC PLAIN (NORTHERN PLAIN)

To the south of Himalaya and north of Peninsular Plateau lies a wide plain composed of recent sediments that were deposited by the rivers flowing from Himalaya comprising those of Indus and Ganga river systems. Known as Indo-Gangetic plain or Northern Plain, the sediments comprising the plain are apparently resting on the northern margin of the peninsular rocks. The Indus system of rivers flows towards the south-west direction into the Arabian Sea while the Ganga river system flows towards the south east into the Bay of Bengal. However, this sedimentary cover is extremely thick, especially around the river valleys. The northern continuation of the Aravalli range separates the plain into the western part formed by the Indus river system and the eastern part formed by the Ganga-Brahmaputra system. The plain extends from Assam in the east to Punjab in the west, extending further into Pakistan. The total length of the plain is approximately 2400 kilometres, while the width is approximately 300 kilometres in the west and approximately 150 kilometres in the east. The plain is a flat area and is the most fertile region of the country.

2.5 DESERT REGION

To the west of the Aravalli chain and south of the Indo-Gangetic plain lies the Thar Desert which extends from Rajasthan through the northern part of Gujarat into the eastern part of Pakistan. It is the ninth largest desert in the world and occupies a substantial area on the north-western margin of peninsular India. The region has arid to semi-arid climate with a maximum rainfall of approximately 150 mm per year. The aeolian (wind borne) sand dunes constitute the main topography of the region. There are three generations of these dunes: the oldest generation constitutes the consolidated deposits that are followed by semi-stable dunes and the youngest active aeolian dunes.The dune sand is mostly derived from alluvium, though there has been some contribution from the Arabian Sea.

The desert was formed due to climate change during Pleistocene times that resulted into the drying of a major river Saraswati which vanished and totally changed the drainage pattern of the area.

2.6 THE ISLANDS

The Islands of India occur in the Bay of Bengal and the Arabian Sea. The islands of the Bay of Bengal numbering over 350 constitute the Andaman-Nicobar group of islands, while those of Arabian Sea numbering well over 43 islands are called Lakshadweep Group. The Andaman and Nicobar Islands extend from north to south in the Bay of Bengal and cover an area of approximately 7950 km^2. Nicobar Islands are to the south of Andamans and comprise only 24 bigger islands. An active volcano is located on one of the Barren Islands in the Andaman-Nicobar group. The Lakshadweep islands are located off the Malabar Coast and constitute an archipelago of 12 atolls and several coral reefs, some of which are partially submerged. The Lakshadweep group covers a land area of approximately 32 km^2, while the total reefal area is over 4200 km^2.

■■■■

3 Peninsular Archean and Proterozoic crystalline rocks

3.1 ARCHEAN AND PROTEROZOIC CRYSTALLINE ROCKS IN INDIA

3.1.1 Problems in the study of Precambrian Crystalline rocks

Precambrian rocks (those formed before Cambrian comprising Archean and Proterozoic rocks) all over the world usually occur in regions that are known as **shield** areas. Shield areas are those regions of the Earth that constitute the nuclei of the continents. They have become tectonically stable after a series of orogenic (mountain building) cycles. Due to these orogenies, the rocks get severely metamorphosed and re-crystallised and are periodically intruded by igneous rocks. Therefore, all shield areas comprise igneous and high-grade metamorphic rocks which are of Archean or Proterozoic age. Some of the more important shield areas of the Earth are the Canadian Shield, Amazonian Shield, Baltic Shield, African Shield, Australian Shield, Antarctic Shield etc. The Indian Shield or what is popularly known as the Deccan Shield (derived from the Sanskrit word *dakshin* meaning south) comprises the Peninsular plateau including its mountain ranges.

Owing to re-crystallisation, high-grade metamorphism and tectonics as well as the fact that the rocks are not stratified sequentially, these rocks cannot be studied through the methods of normal stratigraphic principles. There are usually no fossils in these rocks and, therefore, bio-stratigraphic principles are generally inapplicable. The time control through radioactive dating also does not indicate the time of deposition but that of crystallisation of the rocks, and is usually indicative of the time of orogeny. Thus, these rocks are studied on the basis of their tectonic history.

Tectonically, the shield areas of the world comprise of two main divisions. These divisions constitute the **craton** and **mobile belts**. The term 'craton' is derived from the Greek word meaning 'strength'. Accordingly, cratons are those parts of the shield that are stabilised after a series of cycles of orogeny, metamorphism and igneous activity. As against this the mobile belts are regions in between the cratons that are still tectonically unstable and have suffered severe faulting, fracturing and shearing along its boundaries with the cratons.

3.1.2 Distribution of Precambrian Crystalline Rocks in India

In India the crystalline Precambrian rocks are mainly found in Peninsular India which constitutes one of the important shield areas of the world. These rocks are of Archean and Proterozoic age. While Archean rocks may also be present as the oldest or basement rocks in Himalaya; however, since they are not appropriately dated and the boundary between Archean and Proterozoic is not well defined there, it is not possible in most cases to say whether the rocks occurring there are of Archean or Proterozoic age.

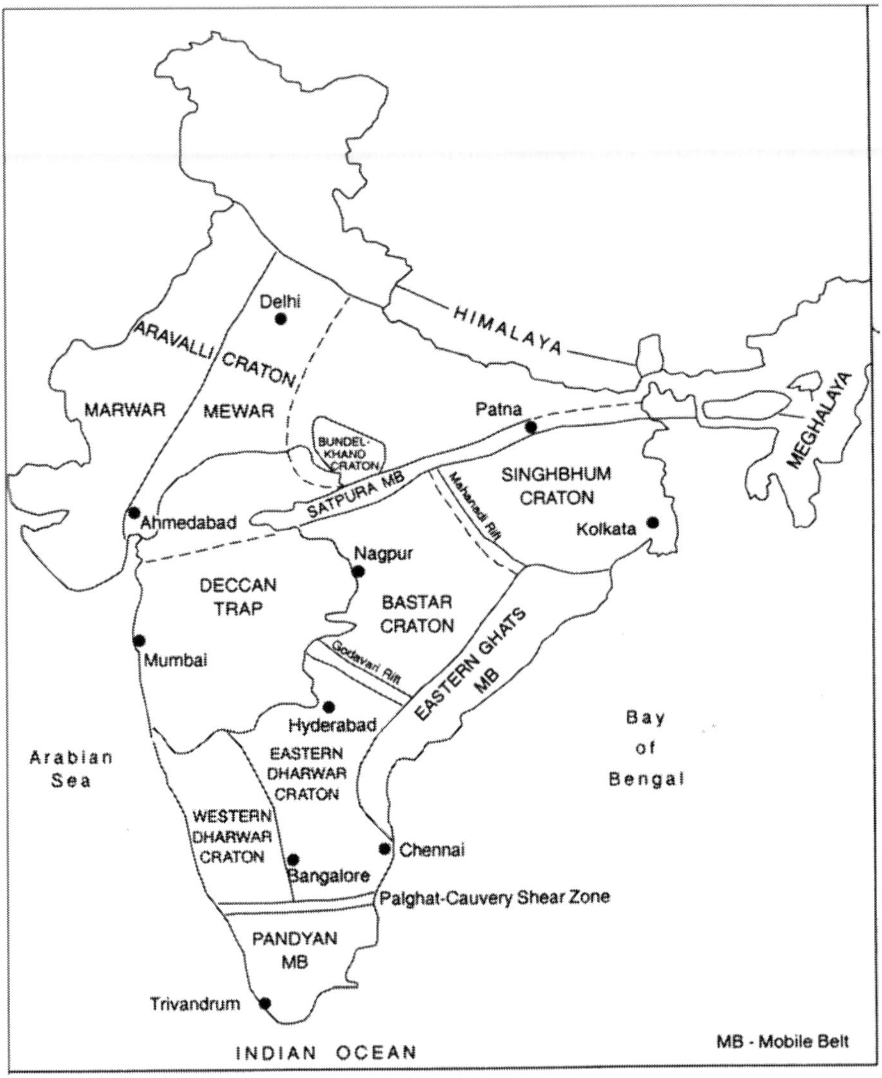

Fig. 3.1. Map of India showing distribution of cratons and mobile belts (After *Geology of India, Vol.* 1 by Ramakrishnan and Vaidyanadhan published by Geological Society of India).

In the Peninsular region the crystalline Precambrian rocks of different types and compositions are found in various cratons and mobile belts. These cratons and mobile belts are distributed all over the peninsular area. The main cratons are Dharwar, Bastar, Singhbhum, Bundhelkhand and Aravalli. The important mobile belts are Eastern Ghat, Pandyan and Satpura (Fig. 3.1).

3.2. DHARWAR CRATON

3.2.1. Location and distribution

The Dharwar craton is located in Karnataka, parts of Tamil Nadu, Telangana and Andhra Pradesh. It is bounded to the south by the Pandyan Mobile Belt, to the east by the Eastern Ghats Mobile Belt, to the north by the Late Cretaceous Deccan Volcanic rocks and Godavari fault. The craton covers an area of approximately 450,000 km^2. The craton is divided into two tectonic blocks—namely, the Western and Eastern Dharwar cratons—both of which are separated by a shear zone. A shear zone is a planar rock area that has undergone higher strain than the adjoining rocks and behaves like a fault system except that the fault plane is not seen anywhere. Another prominent feature separating the two blocks is the occurrence of Closepet Granite massif which runs approximately in the N-S direction. The shear zone separating the two blocks varies in width and is made up of schistose rock. The distribution of the blocks of Dharwar craton in relation to other rocks of India is shown in Fig. 3.2. The Dharwar craton is one of the earliest and best studied cratons of the Peninsular India.

3.2.2. Classification and Lithology

Originally the term Dharwar was applied to schistose rocks in Karnataka that were considered as distinct from the various gneisses and granites (then named as Champion Gneiss, Peninsular Gneiss, Charnockite and Closepet Granite). However, as more detailed work was conducted and geochronological data became available, it was found that the pattern was not as simplistic and the so called Dharwar schistose belt, Peninsular Gneiss and Charnockite had no stratigraphic significance since they did not represent units formed in a sequence. There were similar rocks of varying ages and different rocks of identical ages. Various rock units in the belt crystallised at different points of time and were not classified purely on the basis of lithology.

Moreover, it was observed that the rocks crystallised in the Eastern Dharwar craton were different from those crystallised at the same time in the Western Dharwar craton. The western block of the craton lies to the west of N-S trending Closepet Granite. It comprises the older gneissic complex together with the large schistose belts which constitutes the Dharwar Supergroup. However, the gneissic complex is much older and has been dated from 3,400 to 3,000 Ma. On the basis of geochronological dating, it has been grouped into three stratigraphic units: Gorur Gneiss, Sargur Group and Peninsular Gneiss. These are by far the oldest units of the craton and constitute the basement rocks. Comparatively, the older rocks on the eastern part of Dharwar craton comprise older Basement Gneiss followed by Warangal Group and Peninsular Gneiss.

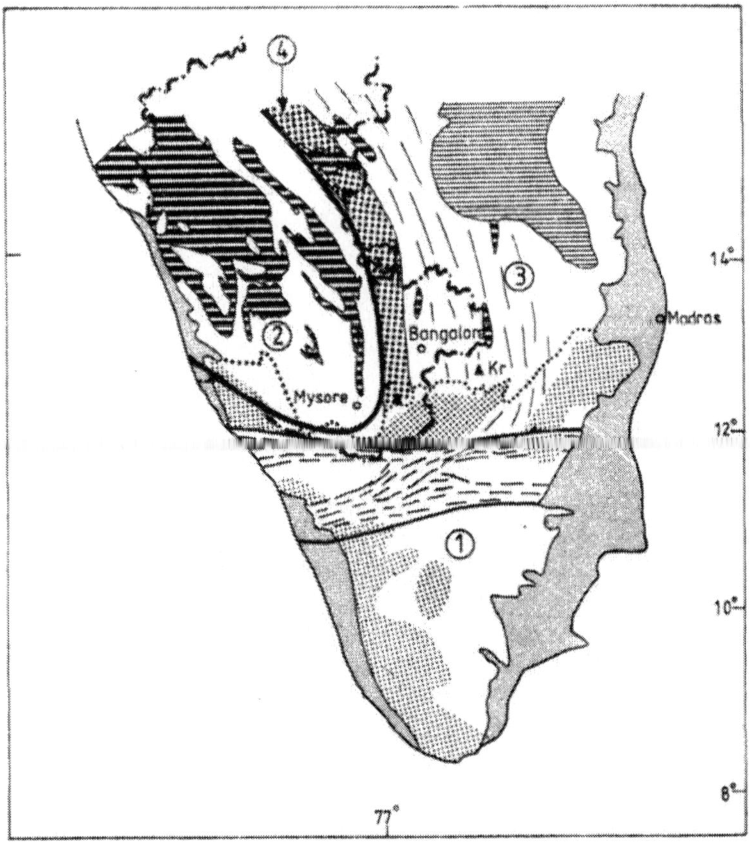

Fig. 3.2. Distribution of blocks of Dharwar craton and the Pandyan mobile belt in South India. 1. Pandian Mobile belt 2. Dharwar western block 3. Dharwar eastern block 4. Closepet Granite separating the eastern and western blocks. (Adapted from *Geology of Karnataka* by Radhakrishna and Vaidyanadhan published by Geological Society of India)

The Dharwar Supergroup that succeeds the above-mentioned basement rocks indicates a geochronological date range of 2800–2600 Ma. This Supergroup is essentially composed of schistose rocks of varying composition. An essential feature of these schistose rocks is that they are dominated by greenstones comprising metamorphosed basic igneous rocks. These rocks include amphobolites, metabasalts and ultramafic schists. The other rocks include low-grade metamorphic derivatives of sedimentary rocks such as graywackes, conglomerates, quartzites and phyllites which are more pronounced in western block. The greenstone rocks are also widespread in the eastern Dharwar block where gold bearing Kolar greenstone belt is located.

An essential element of the schist belts of all cratons of India in general, which includes Dharwar craton, is the occurrence of Banded Iron Formation (BIF). This formation occurs in rocks of varying geochronological ranges between 3100 and 2600 Ma. The striking feature of these rocks is their banded character with layers of chert and jasper alternating with haematite. The occurrence of these rocks broadly at a particular period of time indicates a profuse development of photosynthetic organisms releasing large quantities of oxygen which reacted with iron derived from basic volcanics producing iron ore in the form of haematite. This formation is the major source of iron ore in India.

Another important lithological element of not only the Dharwar craton but many other cratons and mobile belts is a rock type that is popularly known as charnockite (derived from the name of Job Charnock, founder of Kolkata whose tombstone is made of this rock). This is a misnomer since this rock belongs to the granulite facies of metamorphism that is formed under high pressure and temperature conditions. It ranges from acidic to ultramafic composition depending on the nature of the original rock prior to metamorphism. Characteristic feature of this rock is the presence of hypersthene in addition to a wide variety of other minerals. Earlier, it was assumed that this rock unit has a stratigraphic significance. However, geochronological dating has revealed that the crystallisation of charnockites is known at 3000 Ma, 2500 Ma and even 1000 Ma. Evidently, charnockites have been crystallised at different points of time, depending on the grade of metamorphism.

The Dharwar craton is rich in mineral wealth and many minerals are commercially exploited from it. It is more famous for the Kolar and Hutti Gold mines.

A simplified version of the classification and lithology as revised on the basis of recent studies and geochronological data is summarised in Table 3.1.

Table 3.1 Generalised classification of Dharwar craton

Age (in Ma)	Stratigraphic units in Western Dharwar craton	Lithology	Equivalents in Eastern Dharwar craton
2500–2600	Chitaldurga Granite and Charnockite	Biotite granite, granodiorite and charnockite	Closepet Granite and Charnockite
2600–2800	Dharwar Supergroup	Conglomerate, quartzite, Banded iron formation, green schist, amphibolites, volcanics and pyroclasts	Dharwar Supergroup
	----Unconformity---	----------------------------	-----------------
2900 – 3000	Peninsular Gneiss	Gneiss, migmatite, granodiorite	Peninsular Gneiss
3100-3300	Sargur Group	Kyanite-mica schist, garnet schist, amphibolites with ultramafic intrusives	Warangal and Salem Groups
3300-3400	Gorur Gneiss	Gneiss, tonalite, granodiorite	Basement Gneiss

3.3. BASTAR CRATON

3.3.1 Location and distribution

The Bastar craton mostly covers the State of Chhattisgarh and some parts of Andhra Pradesh. It covers an area of approximately 130,000 km^2. It is bounded to the east by the Eastern Ghats mobile belt, to the north-east by Mahanadi River, to the south by river Godavari, to the north by the Satpura mobile belt and to the west by the Deccan Volcanics.

3.3.2 Classification and lithology

The Bastar craton comprises a complex of gneisses, granites, granulites (including charnockites and leptynites), banded iron formation, volcanoclastic rocks and intrusive basic dyke swarms. The craton is a complex that has suffered various orogenic cycles at different points of time and the different orogens have got welded together. Earlier they were classified on the basis of lithological diversity. However, the geochronological studies have helped in identifying the different orogenic cycles, wherein it has revealed that the lithology is not a criterion during crystallisation. One of the prominent complexes in the craton that bears signature of several main orogenies is the Kotli-Dongargarh orogenic complex that gives name to various stratigraphic units. The equivalents to these in other parts of craton are sometimes differently named. The Bastar craton is rich in mineral wealth and bears Malanjkhand copper ore belt and rich iron ore reserves derived from banded iron formation. Table 3.2 gives a generalised classification of the craton.

Table 3.2. Generalised classification of Archean–Proterozoic sequence of Bastar craton

Age (in Ma)	Stratigraphic units in Bastar craton	Lithology
1800-2000	Chilpi Group/Sakoli Group	Phyllites and banded cherts, conglomerate
-2200	Khairagarh Group	Sandstone and basalt alterations
-2300	Dongargarh Granite/ Malanjkhand Granite	Granite to granodiorite (bearing copper ore)
-2300	Nandgaon Group/ Sonakhan Group	Felsic and mafic volcanics, meta basalts and greenstone
-2400	Bailadila Group	Phyllite, quartzite with banded iron formation (rich in iron ore)
2500-2600	Bengpal Group	Sillimanite and andalusite schists with metabasalts
	-----Unconformity----------	---
-2600	Kondagaon Granulites	Charnockites and leptynites

Age (in Ma)	Stratigraphic units in Bastar craton	Lithology
-3000	Amgaon Group/Sukma Group	Kyanite-sillimanite-mica schists, amphibolites, ultramafics with banded iron formation
	-----Unconformity----------	--
3000-3600	Older Gneissic Complex	Grey gneiss

The earliest crustal rocks of this craton comprising gneisses and granodiorites date back to 3500–3600 Ma but their occurrence is scanty. They are followed by widespread enclaves of Sukma and Amgaon rocks of varying composition and grade of metamorphism. The subsequently crystallised rocks comprise andalusite and sillimante schists associated with amphibolites and amygdular metabasalts that are dated around 2600–2500 Ma. Around the same time, there is evidence of large scale intrusive granitic activity. This level marks the beginning of Proterozoic era and the younger rocks are all of the Proterozoic age. This includes the widespread distribution of granites, granodiorites, felsic and mafic volcanics, phyllites and banded iron formation. These and some other younger rock units bear substantial economic deposits of iron ore. Copper and other minerals of commercial value are associated with the same rocks. The youngest elements constitute the metasediments of the Chilpi/Sakoli Groups.

3.4 SINGHBHUM CRATON

3.4.1 Location and distribution

The Singhbhum craton is located in parts of Jharkhand, Orissa and West Bengal. It is bounded in the north by the extension of Satpura mobile belt, in the south by Eastern Ghats mobile belt, in the east by Bengal alluvium and in the west by the Mahanadi rift separating it from the Bastar craton. It is a triangular area covering approximately 50,000 km^2. The core of the craton comprises granite known as Singhbhum Granite bordering which are the different metamorphic and volcanic complexes, representing various orogenic cycles. The southern orogens are high grade metamorphic rocks and older in age which start younging further northwards with the rocks comprising low grade metasediments. A shear zone separates the southern part of Singhbhum craton from the northern part. North of the shear zone the rocks comprise volcanics, cherts, low grade metamorphics and metasediments.

3.4.2 Classification and lithology

The oldest rocks of Singhbhum craton are the Older Metamorphic Group. The rocks of this Group occur in a series of enclaves. These rocks comprise mainly mica schists, quartzites and amphibolites together with an association of a wide variety of other schistose rocks. The amphibolites are derived from the alteration of basalts and retain some of the minerals of the original rock though most of them are altered to amphiboles. The enclaves lie within widespread exposures of granitic rocks that were earlier grouped together as a single unit. However, it has been seen that all the granitic rocks are not of the same age and have been intruded at different points of time. There are various granites of varying ages, as revealed by geochronological dating that were at one time grouped together as the Singhbhum Granite.

Basins bearing banded iron formation are widely distributed bordering the Singhbhum Granite. Banded iron formation is a common rock type in most cratons, as mentioned earlier in the description of Dharwar craton; however, it is widely distributed in Singhbhum craton and is an important source of iron ore. It consists of alternating bands of iron ore (haematite) and quartzite. The formation bearing iron ore in Singhbhum is also older as compared to banded iron formation in other cratons and is believed to be of varying ages that may indicate, in some cases, an age older than 3100 Ma. Dhanjori Group overlying the Singhbhun Granite also bears banded iron formation in addition to metasediments and conglomerates. However, these are of younger ages and correspond to the ages as seen in other cratons.

Table 3.3. Generalised classification of Archean-Proterozoic sequence of Singhbhum craton

Age (in Ma)	Stratigraphic units in Singhbhum craton	Lithology
1100–2000	Newer Dolerite	Comprising three episodes of basic and ultramafic dyke swarms
?1600	Kolhan Group	Conglomerate, sandstone, limestone and shale (Equivalent to Proterozoic sedimentary basins like Cuddapah Group)
1100–2200	Younger Granitoids	Various kinds of granites and granitoids of different ages
?2400	Dalma Group	Mafic and ultramafic lavas comprising basalts and picrites
?2800	Dhanjori Group	Conglomerates and metasediments associated with volcanic rocks and banded iron formation
	---Unconformity—	--
3000–3100	Singhbhum Granite	Three phases of granite and granodiorite, occasionally gneissose in structure
3100–3300	Iron Ore Group (Koira Group)	Banded iron formation
>3300	Older Metamorphic Group	Garnet sillimanite schists, amphibolites

The other common rock types include basic volcanics and greenstone, flyschoidal metasediments, mafic dyke swarms. These are referred to as Dalma Group. The name Dalma was earlier referred to indicate only the volcanic rocks as Dalma 'Trap'. The name is since revised and upgraded to a Group to include the intrusive rocks and associated sediments. There is, however, a difference of opinion about the classification of these rocks. The intrusive rocks include gabbros, anorthosites and ultramafic rocks bearing chromite that constitute the source rock of that mineral which is commercially exploited.

The younger Kolhan Group bears Proterozoic sedimentary rocks that may be equivalent to such rocks as known in Cuddapah basin (see next chapter). These include conglomerates, sandstone, shale and limestone.

A summarised version of the classification based on orogenic cycles as confirmed by geochronology is given in Table 3.3. The craton is rich in mineral wealth including iron ore, chromite and radioactive minerals.

3.5. BUNDHELKHAND CRATON

3.5.1 Location and distribution

The Bundhelkhand craton is northernmost craton of the Indian shield covering parts of states of Madhya Pradesh and Uttar Pradesh. It is bounded to the south by the Satpura mobile belt. To the west it is separated by Vindhyan basin from the Aravalli craton. Towards the north and northeastern side it is covered by the Indo-Gangetic alluvium and may continue under it to further north. This is the smallest craton of Indian shield and the exposed part is approximately 26,000 km^2. Even if the concealed part under the alluvium is considered, it would cover only an area of 35,000 km^2.

3.5.2 Composition and lithology

The craton primarily comprises granite with bands of gneiss and other enclaves of older rocks. In addition, to the north it bears mafic and ultramafic rocks. The granite comprises a large plutonic complex of batholithic size. It consists of porphyritic granite and leucogranite. Intrusives within the main granite are younger leucogranites and syeno-granites. The granite represents two orogenic episodes dating 3200–3300 and 2500–2550 Ma.

The main gneisses indicate an intermediate age of 2700–2900 Ma, while in several other cases they give variable ages. It is possible that they got metamorphosed at different points of time. The gneissose rocks occur in linear belts with an approximate E-W trend. They comprise essentially of banded iron formation interspersed with amphibolites, quartzite and marble. The mafic dykes comprising dolerites and gabbros are a characteristic feature of the granite pluton and they indicate an age of approximately 1800 Ma. Ultramafics are mainly peridotites, dunites and pyroxenites and they are often associated with metabasic rocks.

3.6 ARAVALLI CRATON

3.6.1 Location and distribution

The Aravalli craton extends linearly from north Gujarat cutting across the whole of Rajasthan, parts of Madhya Pradesh and fringes of Haryana, ending up in Delhi. The exposed area of the craton covers approximately 100,000 km^2; however, substantial parts of it are covered to the west by desert sand and to the north and NNE by Indo-Gangetic alluvium. Some isolated exposures in the north within the Indo-Gangetic alluvium indicate that the craton may be extending below that alluvium into the Himalaya. The craton is bounded to the east by the Vindhyan rocks from which the craton is separated by the Great Boundary Fault (GBF). Beyond Vindhyan rocks to the east occurs the Bundelkhand craton discussed above.

As mentioned earlier, the cratonic area is covered by the sand of the Thar Desert to the west. It is bounded to the south by the Narmada rift. The cratonic area is made up of two cratons: the eastern part called Mewar craton and the western part the Marwar craton, the two being separated by a tectonic boundary. The older units of the craton comprising granite and related gneisses are similar to those of the Bundelkhand craton; however, younger elements comprising Proterozoic rocks, which are absent in Bundelkhand craton, are widespread in Aravalli craton. In fact, the main Aravalli mountain range comprises several fold belts of Early and Middle Proterozoic age that were deposited at different points of time.

3.6.2 Classification and lithology

The Mewar craton, comprising the eastern part of the Aravalli mountain range, bears many orogenic cycles ranging from oldest Archean to Meso-Proterozoic. As against that the Marwar craton to the west mainly bears the younger Proterozoic elements comprising the Neo-Proterozoic rocks, including widespread granites and volcanics. The lithology of the two is different and that they are taken together as Aravalli craton is only because they are welded together.

The Mewar craton bears major sets of orogenic cycles namely the Proterozoic Aravalli and Delhi orogenies resting on Archean basement of gneisses. The basic difference between the South Indian cratons and the Aravalli craton is that unlike southern cratons there are no greenstone belts derived from metamorphosed volcanics in Aravalli craton. Instead there is a predominance of quartzites, marbles, pelites and volcanics. In fact, the main marble deposits of India are almost all derived from Aravalli complex.

Banded Gneissic Complex and its equivalent Bhilwara Gneiss constitute the basement units of the Mewar craton. Apart from gneissose rocks they bear amphibolites, quartzites, mica-schists, marbles and lenses of banded iron formation. They are believed to have evolved in two stages. The first stage involves the generation of amphibolites from the mantle material while the second stage involves the partial melting of the amphibolite producing gneisses and granitoids.

The Aravalli Supergroup comprises two rock associations namely, the lower quartzite-carbonate sequence with association of basic volcanics and the upper comprising greywacke-phyllite complex usually referred to as Hindoli Group and sometimes considered as distinct from Aravalli Supergroup. There is a distinct unconformity separating the two. Whether it would be suitable to treat them as a single supergroup is debatable. The main problem about the classification and correlation of the various rock units in the Aravalli Supergroup is because of the partial melting of the basement gneissose rocks and their intrusion within the Aravalli metasediments. The true Aravalli rocks have suffered a much lower grade of metamorphism than the older rocks. As per the radiometric dating, they are Paleoproterozoic in age unlike the basal gneisses (Banded Gneissic Complex) which are undoubtedly Archean. However, the partially molten elements of the complex intruding in Aravalli rocks obviously indicate a younger age and are sometimes confused as a part of the Aravalli complex. The Aravalli rocks are deformed in a series of isoclinal folds due to which their actual thickness is not easy to determine.

Table 3.4. Classification of the Precambrian rocks of Aravalli craton:

Age (in Ma)	Stratigraphic units of Mewar craton	Stratigraphic units of Mewar craton	Lithology	Strati-graphic units of Marwar craton	Lithology
500-550				Marwar Super-group.	Sandstone, shale, dolomite
720-750				Malani igneous suite.	Granite, rhyolite, basalt etc.
800-850				Sirohi Group.	Quartzite, marble Granite
900				Erinpura Granite.	
900-1100		Ajabgarh	Phyllite, Quartzite, marble, volcanics		
	Delhi Supergroup				
1450-1600		Alwar Group	Marble, quartzite, conglomerate, phyllite		
	------------------ Unconformity ------------------				
1700-1800	Hindoli Group		Greywacke, phyllite, quartzite		
	------------------ Unconformity ------------------				
1800-2200	Aravalli Supergroup		Conglomerate, phyllite, meta-basalts, quartzite		
	------------------ Unconformity ------------------				
2600-3300	Banded Gneissic Complex / Bhilwara Supergroup/ Mewar Gneiss		Gneisses and granites		

The younger Delhi Supergroup is dominantly quartzite and marble. Earlier the formation bearing marble and other related rock types including quartzite were separately grouped as Raialo Formation (or Group) with an intermediate position between Aravalli and Delhi Supergroups. However, these rocks are now considered as the basal units of Delhi Supergroup. Delhi Supergroup is divided into Alwar Group and Ajabgarh Group. Alwar Group is largely arenaceous apart from the basal marble rocks mentioned above. As against this Ajabgarh rocks are mainly argillaceous, bulk of it comprising carbonaceous phyllite with subordinate ferruginous sandstone. Delhi Supergroup is mainly of Mesoproterozoic age though some of the younger elements may well be extending into early Neoproterozoic. The Delhi rocks in the Khetri basin and Daraba bear a mineralised copper belt where copper ore is commercially exploited.

The Marwar craton primarily comprises younger rocks mainly of Neoproterozoic age. Its metasedimentary unit comprises a pelite quartzite marble suite known as Sirohi Group, while the igneous suite constitutes the Erinpura Granite. Erinpura Granite is of batholithic dimensions extending for approximately 325 km and ranges from granite to granodiorite. Much younger in age is the Malani igneous suite comprising granites and differentiated volcanics ranging from basalt to rhyolite. These are overlain by sedimentary beds equivalent to Vindhyan sequence (Marwar Supergroup) which will be discussed in the next chapter.

Table 3.4 gives the summarised classification of the Mewar and Marwar cratons of the Aravalli complex.

3. 7 PENINSULAR MOBILE BELTS

3.7.1. Eastern Ghat Mobile Belt (EGMB)

The Eastern Ghat Mobile Belt (EGMB) extends for approximately 1000 km to the east of Dharwar, Bastar and Singhbhum cratons, covering parts of Andhra Pradesh and Orissa. It is a crescent shaped belt tapering in north east and south-west with maximum width in the middle that is approximately 300 km. The belt is overlain in the east by younger Phanerozoic sediments that are bordered by the Bay of Bengal. On the west, it has a tectonic contact with the rocks of the Cuddapah Supergroup and its equivalents bordering Dharwar and Bastar craton. Mahanadi and Godavari grabens cut across the belt.

One of the main rock types that constitute the belt is garnet-sillimanite-graphite gneiss that has been named khondalite (derived from Khond tribe of the region). It is a high grade metamorphic rock belonging to the granulite facies of metamorphism. Associated with it are charnockite, leptynite and migmatites. The belt is divisible longitudinally into four zones: namely, Western Charnockite Zone, Western Khondalite Zone, Central Migmatite Zone and Eastern Khondalite Zone. The radio-active dating of the rocks by various methods indicates a variable age of crystalli-sation with evidences of repeated recrystallisation during various orogenic cycles. The oldest dates are those of charnockites and gneisses of Western Charnockite Zone that give an age of 2800 Ma that is overprinted with age of another cycle around 1500–1700 Ma. The ages of different units are highly variable but there appear to be four main orogenic cycles: namely, 2600–2800 Ma, 1500–1700 Ma, 950–1100 Ma and 800–850 Ma. The last one marks the intrusives of anorthosites, alkaline rocks and granitoids. While earlier it was believed that charnockites represented a particular orogenic event, it has now been conclusively proved, as mentioned earlier in the description of Dharwar craton, that they are of varying ages and as a rock unit do not have any stratigraphic significance.

3.7.2. Pandyan Mobile Belt

The Pandyan Mobile Belt represents the southernmost belt of Indian shield comprising granulite facies of rocks. It is also known as Southern Granulite Terrain (SGT) but that name is somewhat confusing. Actually, rocks belonging to granulite facies are also known in the southern part of Dharwar craton, including charnockites, bordering the Mobile Belt; however, these rocks are of much older age and considerably different from those of the mobile belt. Tectonically, the two also are different; therefore, it is preferable not to use the term SGT for this belt.

The Pandyan Mobile Belt is divided from north to south into three zones: namely, Marginal Zone, Madurai Block and Trivandrum block. The three are separated from each other by shear zones. Marginal Zone borders the Dharwar craton from which it is separated by a shear zone known as Palghat-Cauvery shear zone. Marginal zone bears a wide variety of biotite-hornblende gneisses and migmatites. This zone also bears large massifs comprising charnockites like Nilgiri, Kollimalai and Wayanad. This zone represents the oldest unit of the belt and is comparable and probably derived from the Peninsular Gneiss of Dharwar craton. Other lithological associations of this zone are layered basic complexes comprising gabbros, dunites and pyroxenites as in the case of the Sittampundi and Bhavani Complexes. The age of the emplacement of these complexes is approximately 3000 Ma. However, there are much younger intrusives within the complexes that indicate ages of 2000 and 730 Ma. In this block, there is an increase in the grade of metamorphism southwards from greenschist facies of Dharwar craton to granulite facies culminating in what is called Nilgiri Gneiss which is a charnockitic rock.

Madurai Block occurs to the east of Marginal Zone and is separated from Dharwar craton by the extension of the Palghat-Cauvery shear zone. The rocks comprising the block are of varying compositions and ages. They comprise migmatites, gneisses, quartzites and granulites (including charnockites) and carbonates. These rocks could be recrystallised elements of older rocks since bulk of them indicate a crystallisation age of 560 Ma which corresponds to the youngest orogenic cycle in the peninsular region equivalent to what is known as Pan-African Orogeny. Trivandrum Block is a relatively smaller block comprising various kinds of gneisses interlayered with charnockites.

3.7.3. Satpura Mobile Belt

The Satpura Mobile Belt, is a long E-W trending orogenic belt that divides Indian shield into southern Deccan part (comprising Dharwar, Bastar and Singhbhum cratons) from the northern Bundhelkhand part. This is also a major tectonic division since Son-Narmada, which forms a part of this mobile belt, happens to be an important rift zone comprising several faults. Since the belt is extremely long, there is a lot of variation in lithology and tectonics from west to east.

The middle part of the belt, known as Central Indian Tectonic Zone (CITZ), is bounded by a shear zone to the south and Son-Narmada North Fault to the north. Three main lithological units from north to south—namely, Mahakoshal Group, Betul Group and Sausar Group—interspersed with several granulite belts constitute this zone.The Mahakoshal Group comprises quartzite, carbonate, banded iron formation and mafic volcanics. Granites intruding this Group indicate an age range of crystallisation between 1700 and 1900 Ma. Betul Group is a complex comprising gneisses, quartzites, migmatites, acidic and basic volcanics, carbonates, banded iron formation and mica schists. Older gneisses indicate an age of approximately 1550 Ma, while youngest intrusive granites are in the range of 850 Ma. Sausar Group comprises biotite gneiss, migmatite and charnockite at the base, followed by quartz mica schist and manganese bearing rocks called 'gondite' which is a source of manganese ore that is associated with it. Towards the top it bears marble.

To the east of the CITZ and to the north of Singhbhum craton the Satpura Mobile Belt comprises Chhotanagpur Gneiss followed further eastwards by Shillong Group along the Meghalaya plateau. Consequently, the Satpura Mobile belt becomes the longest mobile belt of the Indian shield and bears a variety of rocks of varying ages and marks an important tectonic zone.

■■■■■

4 Proterozoic Peninsular Sedimentary Basins

4.1. PENINSULAR TOPOGRAPHY DURING PROTEROZOIC

During the Mesoproterozoic and Neoproterozoic the Aravalli mountain chain constituted a major topographic feature of the Peninsular India, which could, in some ways, be compared to the Himalaya of the present day. This mountain chain, striking NE–SW, separated the Great Vindhyan Basin to the east from the smaller western Marwar Basin. Southwards, the Satpura orogen, as discussed earlier, a relatively older and less prominent mountain barrier and probably one of the splayed branches of the Aravalli chain, striking ENE–WSW, separated the Vindhyan Basin to the north from the southern Late Proterozoic basins comprising Chhattisgarh, Cuddapah, Bhima and others. These three main sets of basins were the sites of Proterozoic epicontinental sedimentation in peninsular India[1]. They were separated from each other by the basement high areas which constituted the provenance of their sediments. Of these basins the Vindhyan basin was the largest and was probably continuous to the north with the marginal sea which constitutes the present-day Himalaya. This marginal sea (or *Palaeotethys*) has been a site of sedimentation, not only in Proterozoic but also in greater part of Phanerozoic, though the Vindhyan Basin itself had already been filled up and had emerged by that time. (Fig. 4.1)

[1] These deposits were earlier referred to as the 'Purana' Group. However, that name is not as per stratigraphic nomenclature and, therefore, should not be used.

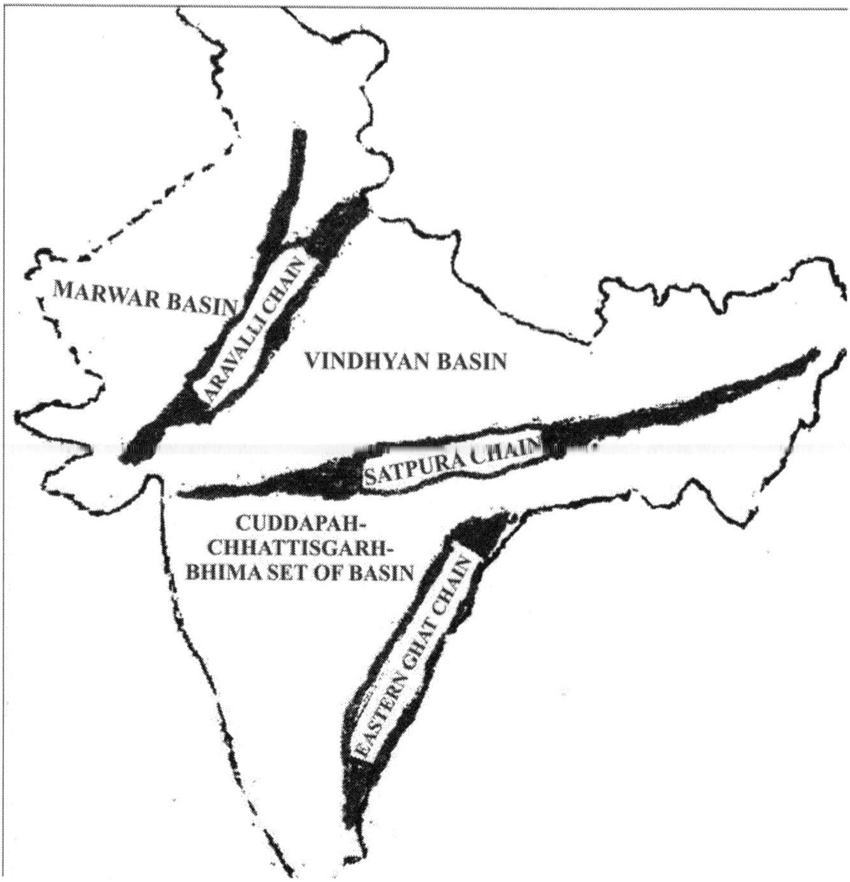

Fig. 4.1. Diagrammatic representation of interpreted topography during Proterozoic

4.2. DISTRIBUTION OF PENINSULAR SEDIMENTARY BASINS

In the Indian Precambrian shield a significant area comprises unmetamorphosed sedimentary rocks of Proterozoic age. These are exposed in series of basins. They cover an area of approximately 150,000 km^2, occupying nearly 20% of the Indian shield. They range in age from 1800 to 600 Ma covering from Late Paleoproterozoic to almost the end of Neoproterozoic. They rest on the crystalline basement with well-marked unconformity or non-conformity. The epicontinental deposition of these sediments is mostly shallow marine but some sedimentary rocks are of fluvial origin. While bulk of the sedimentary basins are mainly of Meso-Neoproterozoic, some Late Paleoproterozoic basins such as Bijawar and Gwalior basins constitute some of the smaller basins. Even some basal units of larger basins like Cuddapah basin may also be of Paleoproterozoic age.

As mentioned earlier, the Vindhyan basin is by far the largest basin; this occurs to the north of the Satpura orogen and the east of Aravalli chain. Adjoining the Vindhyans and Bundhelkhand craton are the smaller basins such as Bijawar, Gwalior and Sonrai.To the west of the Aravalli chain is the Marwar basin comprising Marwar Supergroup (mentioned in Sec. 3.6.2) of Late Neoproterozoic age. To the south of the Satpura chain there are many isolated exposures constituting many basins of varying dimensions. The largest of these are the Cuddapah and Chhattisgarh basins. The relatively smaller ones include Indrawati, Pranhita-Godavari, Bhima, Kaladgi etc.

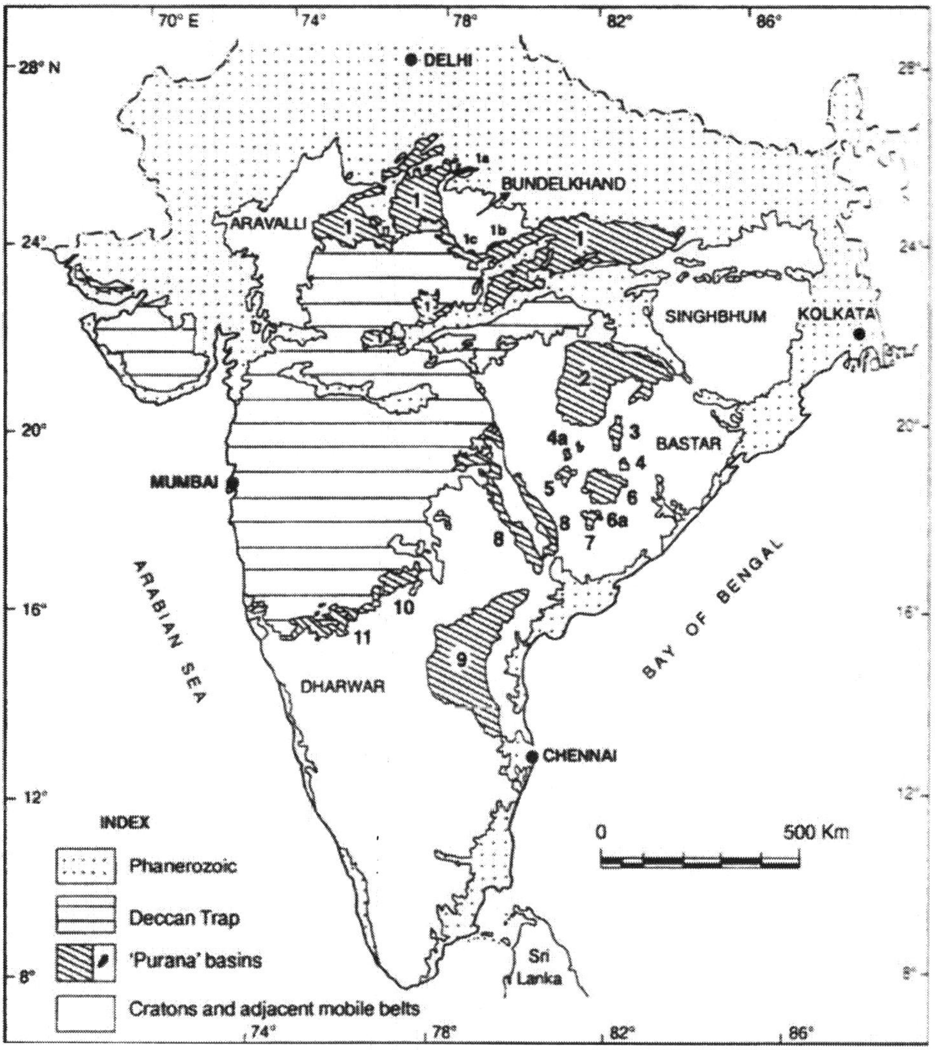

Fig. 4.2. Proterozoic sedimentary basins of Peninsular India (Reproduced from *Geology of India,* Pub. Geological Society of India) 1. Vindhyan, 1a. Gwalior, 1b. Bijawar, 1c Sonrai, 2. Chhattisgarh, 3. Khariar, 4. Ampani, 4a. Keskal and Singanpur outliers, 5. Abujhmar, 6. Indravati, 6a. Chedrapal outlier, 7. Sabari, 8. Pranhita-Godavari, 9. Cuddapah, 10. Bhima, 11. Kaladgi.

4.3. THE VINDHYAN BASIN

4.3.1 Distribution and extent

By far the most important, better-exposed, less-deformed and thickest group of sedimentary rocks in Late Proterozoic is the Vindhyan Supergroup. The name Vindhyan is derived from Vindhyachal mountain range to the north of Narmada River. The Vindhyan rocks extend in a great hatchet-shaped outcrop from near Agra in north-west to Chittorgarh in south-west, Hoshangabad in south and Sasaram in the east. A considerable portion of the rocks is concealed under the Deccan Trap of the Malwa plateau. To the north, the rocks disappear beneath the alluvium of the Ganga basin. The exposures cover an area of approximately 60,000 km^2, while 78,000 km^2 of the rocks are concealed under the Deccan volcanics. The area of the basin covered by the Ganga alluvium is not known but is estimated to be over 84,000 km^2. There is reason to believe that the Vindhyan rocks extend up to the foot of the Himalaya underneath this alluvium and may constitute the basement for the upper Cenozoic Siwalik Group. The rocks are bounded to the west by Aravalli mountain chain from which it is separated by the Great Boundary Fault. The Vindhyan rocks show a total maximum thickness of over 4,000 metres though the individual units may be of varying thicknesses in different sections. The Vindhyan Supergroup deposition extended over a long span of time dating approximately from 1,600 to possibly 600 Ma, though the upper age limit remains doubtful.

4.3.2 Classification and lithology

The Lower Vindhyan rocks are predominantly marine calc-argillaceous sediments, whereas the Upper Vindhyan consists almost entirely of red coloured sandstone and subordinate shale, laid down in paralic environment. There is usually an unconformity between the two and a transgressive overlapping of the red marine or non-marine sediments of Upper Vindhyan over the marine Lower Vindhyan rocks. Table 4.1 gives a summary of the classification and lithological details of the Vindhyan Supergroup.

Semri Group rests with non-conformity on Bundhelkhand Granite or with an unconformity with Bijawar Group. The basal units comprise conglomerate, ferruginous sandstone and shale. Felsic volcanic and porcellanite beds overlie it which is indicative of a volcanic event. Towards the upper part there is the preponderance of limestone and shale. The rocks indicate a varying environment ranging from marine to lagoonal deposition.

Kaimur Group is essentially sandstone-shale association. It also indicates a varying environment of deposition ranging from shallow marine through tidal channel deposit to occasional fresh water stream deposits.

Rewa Group essentially shows deposition in paralic conditions with deeper marine sedimentation at the base shallowing into lagoonal environment with periodic exposure as indicated by mudcracks in the sand-shale sequence. Between Asan Sandstone and Jhiri Shale, there is a diamond bearing conglomerate from which the famous Panna diamonds have been recovered.

Table 4.1. Classification of Vindhyan Supergroup

Divisions	Groups (Thickness)	Formations	Description
Upper	Bhander (1300–1500 m)	Maihar Sandstone Sirbu Shale Bundi Hill Sandstone Lakheri Limestone Ganurgarh Shale	Sandstone, shale with stromatolite bearing limestone. Basal beds are diamond bearing conglomerate
	Rewa (100–300 m)	Govindgarh Sandstone Drummondganj Sandstone Jhiri Shale Asan Sandstone Panna Shale	Glauconite bearing siltstone corsening upwards, Diamondiferous conglomerates between Asan Sandstone and Jhiri shale
	Kaimur (400 m)	Dhandraul Quartzite Mangesar Formation Bijaigarh Shale Markundi Sandstone Ghurma Shale Sasaram Sandstone	Coarse sandstoneinterbedded with shale indicative of lagoonal and intertidal deposition
-----------	----------------	--- Unconformity-----------	------------------------------
Lower	Semri (3000–4000 m)	Suket Shale Rohtas Limestone Chorhat Sandstone Bargawan Limestone Kheinjua Shale Chopan Porcellanite Kajrahat Limestone Arangi Shale Deoland Sandstone	Conglomerate followed by cross-bedded sandstone and shale, porcellanite comprising felsic volcanics; deposition indicative of deeper marine sedimentation.

Bhander shows characteristic limestone-sandstone-shale alteration. Bulk of it is indicative of near shore littoral environment with alternating tidal and aeolean deposition. The basal unit namely Ganurgarh Shale also bears bands of diamond bearing conglomerates as in the case of Rewa Group. Sandstone blocks from this Group have been widely used as building stone for forts and other structures in many parts of India, including Delhi.

4.3.3 Life and age

The Vindhyan rocks bear stromatolites in the calcareous rock units, both in the lower part (Bargawan Limestone) as well as in the upper part (Lakheri Limestone). The

Lower Vindhyan contains *Conophyton cylindricus* and *Collenia columnaris*, while Upper Vindhyan bears *Collenia baicalica.* The former are comparable to Early Mesoproterozoic stromatolites of Siberia, whereas the latter show Middle Mesoproterozoic affinities. The Lakheri Limestone also bears *Obruchevella*, indicating the Late Mesoproterozoic age. The upper part of Lower Vindhyan (Suket Shale) also bears some doubtful algal remains and species of *Chuaria, Tawuia* and *Tasmanites* which have been a subject of debate for over half a century.

Apart from stromatolites a wide variety of microfossils have been obtained from various units of Vindhyan Supergroup but none of these have definite age significance. There are also doubtful reports of small shelly fauna. Ediacara fauna has also been reported from Bhander Group which may indicate its extension upto the end of Proterozoic.

Radiometric dates through K/Ar dating of intrusives and glauconite indicate an age range of 910–1400 Ma for rocks from Semri to Kaimur. Bhander is obviously younger. Because of the presence of Ediacara fauna therein it can be safely presumed that the age of Vindhyan Supergroup would range from the beginning of Mesoproterozoic to the end of Neoproterozoic. There is no evidence to indicate that it extends into Cambrian, as is sometimes claimed.

4.4 THE CUDDAPAH BASIN

4.4.1 Distribution and extent

The Cuddapah basin is the second largest Proterozoic basin of Peninsular India. It is a crescent shaped basin to the east of Deccan plateau with its concave side towards east and convex to the west bordering Dharwar craton. The basin covers an area of approximately 44,000 km^2 and its length along the concave side is approximately 450 km. The average width of the basin is in the neighbourhood of approximately 150 km.The basin actually comprises three sub-basins of varying ages, the oldest Cuddapah *sensu stricto,* Srisailam sub-basin and Kurnool sub-basin. In fact, earlier, the Kurnool sub-basin was considered as an independent basin distinct and younger than Cuddapah basin. However, despite the difference in age, they are now considered as a single basin. The composite thickness of the total sequence ranges from 6,000 to 12,000 m. Its eastern boundary is marked by a thrust that separates it from the Eastern Ghats Mobile Belt. The western boundary is marked by the basic dyke swarms that constitute the basement rocks of the super-group.

4.4.2 Classification and lithology

Cuddapah basin is characterised by quartzite, carbonate and shale occurring in a cyclic order. The basement rocks which are the oldest and probably belong to the upper part of Paleoproterozoic comprise basic volcanics with occasional felsic rocks and tuffs. The lower part of the basin constitutes Cuddapah Supergroup of late Paleopeoterozoic to Mesoproterozoic age while the upper unit comprises Kurnool Group of Neoproterozoic age. A marked unconformity separates the two. Table 4.2

summarises the classification and lithology of the Cuddapah basin including that of the Kurnool Group.

The basal unit of the Cuddapah Supergroup namely Papaghni Group rests with non-conformity on the granitic basement. It consists of quartzite at the base followed by carbonate sequence topped by volcanic rocks. The quartzite formation (Gulcheru Quartzite) also bears conglomerate, chert, jasper and banded iron formation. It is undoubtedly of Palaeoproterozoic age. This is based on the fact that basement igneous rocks on which this group rests have yielded an age of 2100–2400 Ma. The succeeding Vempalle Formation comprises dolostone, siltstone and subordinate sandstone-shale sequence. It is topped by volcanic rocks of basalt-andesite composition.

The overlying Chitravati Group rests disconformably over Papaghani Group and comprises a sequence of quartzite followed by shale-quartzite dolomite alterations. This is interbedded with spilitic pillow lavas indicative of a marine deposition. Stromatolites are also reported from this group.

Nallamalai Group comprises shale-quartzite dolomite alteration which rests on Chitravati Group with a distinct angular unconformity. The different formations in this group have a vertical and lateral passage into each other and have a gradational contact. Only Cumbum Formation can be characteristically identified because of its purely argillaceous nature. Srisailam Quartzite rests on it with a disconformity and comprises glauconitic quartzite.

There is a long time gap between the deposition of Srisailam Quartzite and the overlying Kurnool Group. That is why ealier this group was considered as a different basin than the main Cuddapah basin. However, a long time gap does not indicate that it is a different basin when the deposition is in the same setting. Kurnool deposition had taken place in shallow marine conditions very close to the shore. From conglomerate horizons within the group, placer diamonds have been recovered from time to time. These are derived from kimberlites within and around the basin.

4.4.3 Life and age

Fossils such as *Tawuia* and *Chuaria* from Auk Shale and calcareous algae from Koilkuntla Limestone are not very time specific but broadly indicate a Neoproterozoic age for Kurnool Group. Stromatolites are widely found in different calcareous units of Cuddapah Supergroup but they are not time indicators.

The dykes over which Cuddapah rocks are deposited indicate two ages namely 2400 Ma and 2100 Ma. The dolerite and picrite sills within Pulivendla Quartzite of Chitravati Group indicate an age of 1800 Ma. The other dates available on some younger intrusives and galena dates of Cumbum Shale indicate that post Papaghni part of Cuddapah Supergroup is of Mesoproterozoic age while Kurnool Group is of Neoproterozoic age. All these evidences indicate that Cuddapah Supergroup including Kurnool Group extends from late Paleoproterozoic to Neoproterozoic. However, the upper limit cannot be precisely fixed.

Table 4.2 Classification of Cuddapah Supergroup

Super-group	Group (Thickness in metres)	Formation	Lithology
Cuddapah Supergroup	Kurnool Group (200–450)	Nandyal Shale	Calcareous shale
		Koilkuntla Limestone	Shaly Limestone
		Paniam Quartzite	Quartzite with basal pebble bed
		--------------------------	--------------------------
		Auk Shale	Laminated shale
		Narji Limestone	Limestone with glauconitic Sst.
		Banganapalle Quartzite	Quartzite and conglomerate
	------------	---Unconformity--------	-------------------------------------
	(300)	Srisailam Quartzite	Glauconitic quartzite with shale
	------------	---Unconformity--------	-------------------------------------
	Nallamalai Group (3500–6000)	Cumbum Formation	Slate, phyllite, chert, dolomite
		Pullampet Formation	Shale, barites, dolomite, tuff
		Bairenkonda Quartzite	Quartzite, shale
		Nagari Quartzite	Quartzite, shale, conglomerate
	------------	---Unconformity--------	-------------------------------------
	Chitravati Group (4900–5000)	Gandikota Quartzite	Shale, glauconitic quartzite
		Tadpatri Formation	Shale, tuff, stromatolitic dolomite
		Pulivendla Quartzite	Quartzite, conglomerate
	Papaghni Group (2100)	Vempalle Formation	Stromatolitic dolomite, basic lava
		Gulcheru Quartzite	Quartzite, conglomerate

4.5 THE CHHATTISGARH BASIN

4.5.1 Distribution and extent

Chhattisgarh is the third largest basin of Proterozoic rocks of Peninsular India and covers an area of approximately 36,000 km^2. Its southern boundary is marked by the Bastar craton, whereas it is bounded by the Satpura mobile belt to the north. It has an irregular rectangular configuration with an E-W length of approximately 300 kilometres and N-S width of approximately 200 kilometres. While the total thickness of the rocks does not exceed 2,700 m. it contains some of the thickest deposits of limestone that are of considerable commercial use in cement and steel industry.

4.5.2 Classification and lithology

The rocks of the basin rest on gneisses, greenstones and other crystalline rocks of Bastar craton. This basin comprises three groups of rocks namely Singhora Group, Chandrapur Group and Raipur Group. The lower Singhora Group comprises primarily of conglomerate, arkose and sandstone with frequent intercalations of shale. The middle Chandrapur Group is primarily composed of sandstone resting on basal conglomerate. The upper Raipur Group is mostly made up of carbonate rocks comprising limestone, dolostone with occasional bands of gypsum. Basic dykes intrude the sequence at different levels.

4.5.3 Life and age

The upper part bears a widespread distribution of stromatolites. They comprise two different assemblages belonging to Middle and Upper Mesoproterozoic age. Besides the other biota include prokaryotic filamentous cyanobacteria and acritarchs. The radioactive age is doubtful but the zircon age of ignimbrites in the upper part (Gunderdehi Shale) of the sequence shows an age between 990 and 1020 Ma. The total range of the basin could be from Mesoproterozoic to early Neoproterozoic. Table 4.3 gives a summary of the sequence of this basin.

Table 4.3. Classification of Chhattisgarh Supergroup

Super-Group	Group (Thickness)	Formation	Lithology
Chhattisgarh	Raipur Group	Maniari Shale	Purple shale, dolostone
	(1900 m)	Hirri Dolomite	Grey dolostone
		Tarenga Shale	Shale. Limestone, chert
		Chandi Limestone	Stromatolitic dolostone
		Gunderdehi Shale	Shale, ignimbrite
		Charmuria Limestone	Phosphatic limestone
		Bijepar Shale	Calcareous shale
		Kansapathar Sandstone	Glauconitic sandstone
	Chandrapur Group	Chaporadih Shale	Shale, arenite
	(400 m)	Lohardih Conglomerate	Arkose, conglomerate
	---------------------	--Unconformity-------	-----------------------------
		Chhuipali Shale	Shale, dolostone
	Singhora Group	Bhalukona Sandstone	Sandstone
	(400 m.)	Saraipali Shale	Shale, porcellanite
		Rhatikhol Conglomerate	Conglomerate, arkose, arenite

4.6 OTHER PROTEROZOIC BASINS

4.6.1 Basins to the north of Satpura orogen

As mentioned in Section 4.2 there are a large number of smaller Proterozoic basins distributed throughout Peninsular India. Many of them are remnants of what must have been continuous basins that got separated due to subsequent uplift and erosion. The location of all these basins together with the already described three large basins is shown in Fig. 4.2. Of these there are very few basins to the north of Satpura mobile belt. These are by and large small basins surrounding the Bundhelkhand craton. They include the Bijawar, Gwalior and Sonrai basins.

Bijawar Basin lies between granite of the Bundhelkhand craton and Vindhyan sediments resting with non-conformity on the granite. It has a maximum estimated thickness of 1000 m. It commences with conglomerate followed by basalt flows and stromatolite bearing dolostone. The upper part of the sequence consists of phosphorite bearing sandstone and breccia. The radioactive data as obtained from the basic lavas indicates an age range from 1700 to 1800 Ma.

Gwalior Basin lies 200 km. to the northwest of Bijawar basin and to the north of Bundhelkhand Granite over which the basin is resting. The lithology comprises conglomerate, glauconitic sandstone and stromatolitic limestone. Towards the upper part shale, chert and jasper are associated with basalts and gabbro sills. The age range as indicated from radioactive data of basalts is the same as that of Bijawar basin.

Sonrai Basin lies to the west of Bijawar basin. The lithology comprises grit shale and dolostone, Pillow basalts overlie the sediments. Sonrai basin is considered as an extension of the Bijawar basin that got separated from the latter through tectonic activity.

4.6.2 Basins to the south of Satpura orogen

There are large number of Proterozoic sedimentary basins to the south of Satpura and Narmada valley in addition to the already described Cuddapah and Chhattisgarh basins. These are of varying ages ranging from Mesoproterozoic to Neoproterozoic. The more important ones include Indravati, Pranhita-Godavari, Khariar, Ampani, Abujhmar, Sabari, Bhima and Kaladgi basins. Many of them namely Indravati, Khariar, Ampani, Abujhmar and Sabari surround or rest on the Bastar craton and bear more or less identical lithology comprising quartzite – shale – carbonate sequence. They are detached outliers of what must have been at one time a continuous basin. There are several other smaller detached outliers resting on the craton. The age range is from lare Paleoproterozoic to Neoproterozoic.

Pranhita-Godavari basin comprises several outliers along the Godavari valley at the junction between Dharwar and Bastar cratons. The more notable are the eastern and western belts or sub-basins separated by overlying Gondwana sediments. The sequence is divided in four Groups namely Mallapalli, Mulug, Penganga and Sullavai. The lithology comprises quartzite-dolostone-sandstone. The estimated ages on the basis of glauconite bearing rocks are in the range of 1500–1600 Ma.

Bhima Basin lies to the north-western side of Cuddapah-Kurnool basin. It rests on the Peninsular Gneiss of Dharwar craton with a marked unconformity and is in turn covered by Deccan Trap to the north. The width of the outcrops is variable with a maximum (approximately 40 km) along the Bhima River from which the basin derives its name. The maximum thickness is approximately 270 metres. The lithology comprises grit with thick limestone deposits towards the upper part. The lower clastic rocks appear to be deposited in shallow marine environment while the upper limestone was deposited in a protected tidal flat environment. The rocks are possible equivalents of Kurnool Group and, therefore, could be of Neoproterozoic age.

Kaladgi Basin is one of the larger basins resting on Peninsular Gneiss of Dharwar craton covering an area of approximately 8300 km^2. It is exposed principally in Belgaun, Bijapur, Dharwar and Gulbarga districts. It comprises two sequences, the older Kaladgi and younger Badami sequence. The older Kaladgi sequence referred to as Bagalkot Group comprises breccias-quartzite- dolostone sequence and is equivalent to older Cuddapah rocks. It is overlain by Badami Group with a distinct angular unconformity. Badami Group has a gentle dip and comprises essentially arenaceous rocks and is equivalent to Kurnool Group. The dolostone horizons of Bagalkot Group show a profuse development of stromatolites namely *Colonella, Conophyton* and *Collenia.*

4.6.3 Basin to the west of Aravalli chain

To the west of Aravalli chain in the north lie several outcrops of what is referred to as West Marwar Basin. The trans-Aravalli region bears numerous small outcrops of sedimentary rocks associated with the Malani suite of igneous rocks, but the bulk of these rocks is concealed under the desert sand. In lithology and sedimentary structures, the rocks are comparable to those of Upper Vindhyan (Bhander). The rocks outcrop from Jodhpur to Pokharan in Rajasthan (Jodhpur Sandstone) which rest on the eroded surfaces of Malani volcanics (already referred to in Table 3.4). The latter represent extrusive phase of a suite of felsic igneous rocks of which the intrusive suite comprises granitic rocks (Siwana and Jalor Granites). The radiometric dating of these igneous rocks indicate that they are co-eval with Upper Vindhyan rocks since they do not give dates older than 940 Ma. Since the Jodhpur Sandstone non-conformably overlies these rocks, it could be equivalent to the youngest units of Vindhyan sequence. In Nagaur area gypsiferous rocks associated with dolomite are exposed within the desert sand. These rocks are presumed to be younger than the Jodhpur Sandstone and may be equivalent to the evaporite facies (Saline 'Series') of Salt Range of Pakistan. The northerly extension of Marwar basin is not known since most of it is covered by Thar desert sand and the Jurassic rocks of Jaisalmer.

5 Precambrian Rocks of Himalaya

5.1 CRYSTALLINE ROCKS

5.1.1 Significance of Crystalline Rocks in Himalaya

There is a basic difference between crystalline rocks of peninsular India and those of Himalaya. The crystalline rocks of peninsula are extremely old and show imprints of several orogenic cycles ranging from rocks as old as those of 3300 Ma. The different orogenic cycles in peninsular India are appropriately dated and it is possible to determine the sequence of these cycles. As against that the crystalline rocks in Himalaya have been overprinted by the relatively younger Himalayan orogeny. Thus, many crystalline rocks of Himalaya are of much younger ages while some of them are remobilised old rocks. Moreover, most of the crystalline rocks of Himalaya are not appropriately dated as in the case of those of peninsular India. In the case of Precambrian rocks of Himalaya, it is difficult to determine if they are of Archean or Proterozoic ages. Accordingly, the rocks are classified on the degree of metamorphism. Thus, the crystalline rocks are separately addressed as distinct from less metamorphosed rocks. However, it needs be mentioned that this does not necessarily indicate that all crystalline rocks are older in age than the less-metamorphosed rocks.

5.1.2 Distribution of crystalline rocks in Himalaya

Crystalline rocks are present in both Higher Himalaya (Himadri) and Lesser Himalaya. In Higher Himalaya, they form the basement complex of gneisses, schists and granites that extend throughout the range from Nanga Parbat in the west through Tso Morari in Ladakh, Dhauladhar and Rohtang complex in Himachal Pradesh, higher ranges of Uttarakhand through Badrinath and Munshiari, higher reaches of Nepal, to Darjeeling, Sikkim, Bhutan and Arunachal Pradesh in the east. They are variously named in different places. The more commonly used names include Salkhala Group in Kashmir, Tso Morari Group in Ladakh, Rohtang Group in parts of Himachal, Vaikrita Group in Himachal and Uttarakhand, Munshiari Formation in parts of Uttarakhand, Himal Group in Nepal, Kanchanjhunga Gneiss in Sikkim, Thimpu Group in Bhutan and Se La Group in Arunachal Pradesh.

In Lesser Himalaya, the crystalline rocks are not continuously exposed as in the case of Higher Himalaya but occur mostly as isolated outcrops as thrust sheets or klippes. Thrust sheets include those of Baijnath, Almora, Askot and Daramgarh in Himachal Pradesh and Uttarakhand. The more important outcrops as windows include the Rampur–Larji window in Himachal Pradesh. Crystalline rocks are also found in Lesser Himalaya of Nepal, Darjeeling area and Bhutan.

5.1.3 Lithology of crystalline rocks of Higher Himalaya

The crystalline rocks of Higher Himalaya are generally referred to as Central Crystallines especially in Himachal Pradesh and Uttarakhand. They comprise essentially of gneisses and schists showing varying degrees of metamorphism and are frequently intruded by younger granites. Higher-grade metamorphics include sillimanite gneisses, followed by kyanite and garnet staurolite schists and lower grade mica schists and occasional quartzites and marbles.

Nanga Parbat crystallines comprising gneisses belonging to amphibolite facies represent some of the oldest dated rocks in Himalaya where zircons have given an age range of 1800–2200 Ma. There are also evidences of vestiges of older Neoarchean rocks that have been remobilised and overprinted by younger rocks. However, extremely young tourmaline granites dating 10–18 Ma intrude the older rocks at various places making it a complex massif. Nanga Parbat Crystallines continue eastwards in Kishanganga valley to Burzula pass in Gilgit.

Tso Morari Crystalline complex in Ladakh constitutes a dome comprising gneisses, eclogites and metasediments intruded by granites. While the gneisses could be of older age, younger granites intrude the complex that give dates as young as 460 Ma that is correlated with Pan-African orogeny that affected many parts of Himalayan region.

Rohtang and Vaikrita Crystallines of Himachal Pradesh comprise garnet–staurolite–kyanite–sillimanite gneisses in the lower part and biotite schists and amphibolites in the upper part. Wangtu Gneiss indicates a Rb–Sr age of approximately 2025 Ma. There are several granite and granodiorite intrusions dating approximately 580 Ma that can be assigned to Pan-African orogeny.

In Uttarakhand, the Crystallines extend all along the Higher Himalayan Range constituting the water divide. They range from sillimanite gneisses through kyanite schist to quartzites and mica schists. The ages of different units range from1840 to 2120 Ma.

In Eastern Himalaya, the Darjeeling Gneiss is essentially kyanite-sillimanite gneiss, while the Kanchajhunga Gneiss is generally migmatite or high grade metamorphics. The Se La Group in Arunachal comprises a complex of sillimanite gneiss, amphibolites, graphite schist and quartzite.

5.1.4 Lithology of Crystalline Rocks of Lesser Himalaya

As mentioned earlier, the crystalline rocks of Lesser Himalaya occur in isolated units as thrust sheets and windows. The thrust sheets of Baijnath–Almora–Askot–

Darangarh comprise graphite schists, quartzite, marble and augen gneiss. The Kishtwar and Rampur Larji windows bear a lithology of migmatitic gneiss, mica schist, quartzite and marble.

5.2 PROTEROZOIC BASINS OF HIMALAYAN REGION

5.2.1 Nature of Deposition

The Proterozoic sequence in the Himalayan area is more difficult to resolve than that of the peninsula because the rocks have been involved in Himalayan orogeny, resulting in severe folding, thrusting and other structural deformations. The order of superposition is disturbed because of these structural complications. In the absence of organic remains which could establish the appropriate stratigraphic sequence, the classification and correlation becomes arbitrary and subjective. This is especially true of the rocks of Lesser Himalaya which are severely thrusted and bear several nappes, involving different stratigraphic units. As against this, the Proterozoic rocks in the Higher Himalaya ('Tethys' facies) are relatively less deformed and are stratigraphically underlying the fossiliferous Phanerozoic rocks. The two regions represent different facies of rocks, and although they must have been deposited in the same basin, the Lesser Himalayan rocks represent the epicontinental deposition on the inner shelf. They would be, therefore, comparable to the deposition in the Vindhyan basin or its western equivalent. As against this, the rocks of the Higher Himalaya were deposited in the outer shelf or beyond and are, therefore, considerably distinct from the Lesser Himalayan facies.

5.2.2 Proterozoic Rocks of Lesser Himalaya

The stratigraphic resolution of Proterozoic rocks of Lesser Himalaya is beset with problems since superposition is generally disturbed due to structural complications. The different facies of rocks cannot be correlated satisfactorily. The classification is, therefore, more on the lithofacies concept and time correlation is only subjective, except where the superposition is undisturbed. (For distribution of Late Proterozoic basins in western Himalaya refer to Fig. 7.1).

The early Late Proterozoic rocks can be divided into following three distinct facies:

1. *Calcareous facies* (Jammu, Shali, Deoban, Tejam, Nawakot, Baxa suite of rocks): This group comprises essentially dolomitic limestone associated with quartzite and slate. They are bearing stromatolites of Middle Mesoproterozoic affinities viz. *Collenia columnaris, C. baicalica, C. symmetrica* etc. These rock units could accordingly be correlated to Upper Vindhyan Bhander Limestone which bears identical stromatolitic forms. Lead isotope age of galena from Jammu Limestone (also known as Sirban Limestone) indicates an age of approximately 900 Ma, which is comparable to uppermost limit of Vindhyan Supergroup. However, since galena is apparently of epigenetic origin, it would indicate a younger age than that of the limestone. This limestone facies extends throughout Lesser Himalaya

from west (Jammu Limestone) to east (Baxa Dolomite) and is probably the only facies which can be correlated with some degree of certainty.

2. *Argillaceous facies* (Simla, Jaunsar, Chandpur suite of rocks): This facies is possibly younger than the calcareous facies, though its being a lateral variant, at least towards the upper part of the latter, cannot be ruled out. The facies is, however, distinctly flyschoidal since it contains rhythmites of shale-siltstone-limestone and greywackes.
3. *Arenaceous facies* (Chail, Tanawal, Nagthat suite of rocks): This group of rocks is variously considered as a facies variant of the Simla Group or younger in age. The mutual relationship cannot be determined since the contact is always tectonic. The facies is essentially arenaceous, with a considerable component of quartzite. These rocks in western Kashmir and Kumaun are associated with basic volcanics (Bafliaz and Bhimtal) and this may represent bimodal volcanism but essentially of the island arc type.

The youngest Proterozoic rocks in Lesser Himalaya are known from the Krol belt and comprise the following facies:

1. *Tillite facies* (Blaini Group): In addition to shale, this group contains significant diamictite beds which have been variously regarded as glacio-marine (tillite) or tectonic (tilloid) in origin. Therefore, it had been, from time to time, correlated to Permo-Carboniferous glaciation of Gondwana-land. However, the conclusive faunal evidence regarding Early Cambrian age for topmost units of Krol Group has finally established its Late Proterozoic age. It represents a phase of a global glacial event referred popularly as 'Snowball' earth.
2. *Carbonate facies* (Krol Group): Overlying the Blaini Group in the Krol belt is the carbonate-red shale sequence of Krol Group. This unit had earlier been given varying ages ranging from Permian to Jurassic, mostly on the basis of fauna which is doubtful or spurious. The controversy has since been resolved because overlying the topmost unit of Krol, chert-phosphorite-black shale unit of the Tal Formation bears earliest Cambrian (comparable to Meischcunian of China) small shelly fauna.

The lower units of Krol Formation, being essentially carbonate in composition, are also sometimes correlated with older calcareous facies (Deoban–Tejam Group) because they also bear stromatolites. However, these stromatolites are of a different type and this correlation does not seem to be valid. Neither is the succession identical nor does the lithology correspond with it.

5.2.3 Late Proterozoic Rocks of Higher Himalaya

In Higher Himalaya, the fossiliferous Phanerozoic rocks ('Tethys' facies) rest on Late Proterozoic rocks and the contact is a transitional one. The Higher Himalaya comprising crystalline rocks mentioned above separates the Lesser Himalaya from 'Tethys' Himalaya. Part of this Higher Himalayan Crystalline Complex constitutes the basement for the Late Proterozoic sequence, though some younger rocks are also involved in the metamorphism of the Crystallines, which has taken place in Cenozoic times.

The Late Proterozoic rocks are differently named in different parts of Higher Himalaya (Dogra Slate or Machhal and Lolab Formations in Kashmir, Haimanta or Batal Formations in Spiti-Zanskar area, part of Martoli Formation in Kumaun etc.). All these formations share almost a uniform lithology in that they are essentially argillaceous with subordinate arenaceous elements. There is no carbonate facies in these rocks, and accordingly, no stromatolites comparable to those of Lesser Himalaya or Upper Vindhyans are found. Age-wise, these formations can probably be correlated to Blaini–Krol Group of Lesser Himalaya. However, they represent an entirely different facies and the deposition seems to have taken place in the outer shelf. The sedimentation is continuous without any apparent breaks. Except for a wide variety of trace fossils and some doubtful cryptarch algal remains of Late Mesoproterozoic affinities, no other fossils are known from these rocks. In almost all areas, these rocks pass conformably upwards into fossiliferous Lower Cambrian.

5.3 PALAEOGEOGRAPHY OF THE INDIAN SUBCONTINENT IN LATE PROTEROZOIC

5.3.1 Ediacaran Sedimentation

If we look at an overall picture of the Indian subcontinent towards the later part of Proterozoic, we notice that in peninsular India the sedimentation came to an end by the middle of Neoproterozoic, which corresponds to approximately 800 Ma. All the peninsular basins had by then been filled up and the subsequent sedimentation was restricted to the extra-peninsular basin alone. Accordingly, despite thorough investigations, no organic remains of definite Ediacaran age are known anywhere in the peninsular rocks, although they are reported in the Himalayan sequence. The only doubtful ones are those reported from Marwar basin of western Rajasthan overlying the Jodhpur Formation.

By the beginning of Ediacaran times, the peninsula had already attained its present configuration and from Ediacaran onwards, and throughout the Phanerozoic, the only marine sedimentation in peninsular India is a result of transgressions and restricted to coastal regions alone. This is also true of Lesser Himalaya which represented an inner shelf of the Himalayan Tethys. The Proterozoic sedimentation in this part corresponds to that of peninsular India, up to the end of Mesoproterozoic. However, after Mesoproterozoic, there is no sedimentation in peninsular India, but the sedimentation continued in Lesser Himalaya (Blaini–Krol–Tal sequence) through Ediacaran upto Early Cambrian.

5.3.2 Proto-Gondwana

The palaeomagnetic data accumulated during last three decades suggests that the Indian shield towards the end of Proterozoic and the dawn of Cambrian, constituted a part of an aggregation of continents including present-day Iran, Afghanistan, China, Tibet, Africa, South America and Australia. (see Sec. 14.1 for details). This landmass was the proto-Gondwana and it was located in the tropical region of the globe. This is one of the reasons that there is a distribution of evaporite facies, though of varying ages [Saline 'Series' of Salt Range (Pakistan) and Mandi, Hormuz Formation of Iran] towards the end of Late Proterozoic. This would also be coeval with a phase of regression at the start of Neoproterozoic, when sedimentation ended in peninsular India. The proto-Gondwana craton was dissected by a network of linear epicontinental seas which included the Himalayan Prototethys. These seas were the sites of sedimentation even during the regressive cycle, which accounts for continuous sedimentation in Higher Himalaya. After Mesoproterozoic, a greater portion of this landmass seems to have suffered a phase of glaciation during Varanger times (Snowball event). It is generally believed that the melting of these glaciers resulted in the deposition of tillites in Lesser Himalaya (Blaini Group; refer to section 5.2.2).

5.3.3 Ediacaran Transgression

The later part of Ediacaran marked a global transgressive phase. This followed the Varanger glacial episode. The transgressive phase, which persisted into Cambrian, left its imprint in the Himalayan sequence, where this transgressive event is recorded both in Lesser Himalaya and in Higher Himalaya ('Tethys' facies). In the Lesser Himalayan sequence (Blaini–Krol–Tal) the transgressive event is limited and extends only up to Early Cambrian. Most of this sedimentation was of a coastal type and was, therefore, marked by a series of smaller transgressive-regressive cycles, resulting in several para-conformities and rapid facies changes. As against this, in Higher Himalaya, which constituted a deeper part of the shelf, the transgressive event extended up to the beginning of Late Cambrian. Moreover, the smaller transgressive-regressive cycles do not leave any appreciable imprint on the sedimentation and there are no breaks, except for a few minor diastems. The transgressive phase, however, did not extend into peninsular India and consequently no sedimentation took place there during this period.

6 Phanerozoic Sedimentation and Transgressive Events

6.1 PATTERN OF SEDIMENTATION

6.1.1 Marine Sedimentation

As discussed in the previous chapter, the marine deposition through greater part of the Phanerozoic time is mostly restricted to Himalayan region. It is so because the Tethys[1] was the main site of sedimentation and the peninsular basins were already filled up and had become areas of positive relief. However, during major phases of transgression, most of the coastal areas of the peninsula got inundated and became sites of marine and paralic sedimentation. In addition, following the various events of diastrophism, fresh water sedimentary basins were formed from time to time in the peninsular region.

It was believed for a long time that sedimentation in the Himalayan Tethys was continuous from Cambrian to Cretaceous. This was also a corollary to the then concept that Himalayan Tethys represented a 'geosyncline', which was continuously sinking and, therefore, received sediments throughout Paleozoic and Mesozoic. The Tethyan sedimentation has now been proved to be of platform type, and like all platform deposits, there are several breaks, some of them extending for extremely long periods. Moreover, the eustatic changes and the instability of the shelf margin resulted in periodic emergence of some regions and their consequent erosion. Most of these breaks cannot be physically identified in the sequence as they are of the

---

[1] The term 'Tethys' was originally applied to a continuous linear sea/ocean which separated Gondwana landmass from Laurasia and came into existence in Late Carboniferous and continued through Mesozoic. In its original connotation, the Tethys was continuous from Andes to Himalaya and included the Alpine basin. The application of the term subsequently became more generalised, and since a linear sea existed in Himalaya from Late Precambrian times, it was also termed as Tethys. It is preferable to use the term 'Tethys' in a broad sense to denote the marine basin from Precambrian onwards and apply the term 'Palaeotethys' for the sea from Precambrian to Carboniferous and 'Neotethys' from Carboniferous to Cretaceous.

nature of disconformities or paraconformities. However, the breaks can be located through faunal record. Seen in this background, there is not much of a difference between the marine coastal facies in peninsular India and the Tethyan facies of Himalaya. It is primarily a difference of depth at which sedimentation has taken place. During a greater part of Paleozoic and Mesozoic, there was deposition going on in Tethys, while in the coastal regions, for the most part, there was no deposition or only erosion. The major transgressive events that inundated the Higher Himalaya also inundated the coastal regions of Lesser Himalaya and peninsular India. The smaller transgressive events are not recorded in either of them and are only recorded in Higher Himalaya. Hence, while all the transgressive events are recorded in Higher Himalaya and have resulted in sedimentation, it is only the major transgressions which have resulted in sedimentation in Lesser Himalaya or the peninsular coastal belt. This is especially true of Late Precambrian and Paleozoic transgressions. For Mesozoic transgression, the pattern is somewhat different. For the main transgressive and regressive phases and corresponding sedimentation in Tethys Himalaya, Lesser Himalaya and peninsular India, refer to Fig. 6.1.

In the Himalayan region, major facies changes were taking place during Cenozoic due to orogeny and closure of Tethys. Accordingly, the initial shallow marine sedimentation in the outer Himalaya and trans-Himalaya got progressively replaced by fresh water sedimentation, as the sea finally receded. The transgressive-regressive cyclicity, as seen in Paleozoic and Mesozoic times, was not operative in Cenozoic, since the sedimentation was no longer on a stable platform.

6.1.2 Fresh Water Sedimentation

While the marine sedimentation in Paleozoic and Mesozoic is primarily restricted to Higher Himalaya and secondarily to the Lesser Himalaya and coastal belts of peninsular India, the fresh water sedimentation during these two eras is mostly restricted to peninsular India. Many of these fresh water basins were produced due to epeirogenic activity and rifting through plate movement as is the case with the Gondwana basins. There was no sedimentation in peninsular India after Vindhyan and equivalent deposition, up to the beginning of Permian. In Early Permian, the Gondwana basins came into existence and received sediments following the melting of glaciers. These basins remained active sites of deposition through Permian and almost the entire Mesozoic when further diastrophic activity in Cretaceous ultimately terminated this sedimentation.

The pattern of fresh water sedimentation in Late Cretaceous (Lameta and intertrappean beds) and Cenozoic in peninsular India is somewhat different from that of Gondwana in that the basins were localised and had restricted time spans. These basins of sedimentation existed in Late Cretaceous, Neogene and Quaternary times.

In the Himalayan region, the fresh water sedimentation is almost restricted to later part of Cenozoic, when Tethys had closed and the Himalaya acquired some amount of relief. A linear fresh water basin (Siwalik basin) was formed in front of the newly emerging orogen and was the site of sedimentation of the molasse facies. Prior to this, sedimentation of the paralic type was taking place (Murree, Dagshai) before the sea receded completely. In the central region, the sea receded in the Oligocene times, whereas in the eastern (Assam, Meghalaya) and western (Sind–Baluchistan) regions it continued well into Miocene. Accordingly, there is a lateral transition from fresh water through paralic to shallow marine sedimentation during Palaeogene and part of Early Neogene in outer Himalayan region.

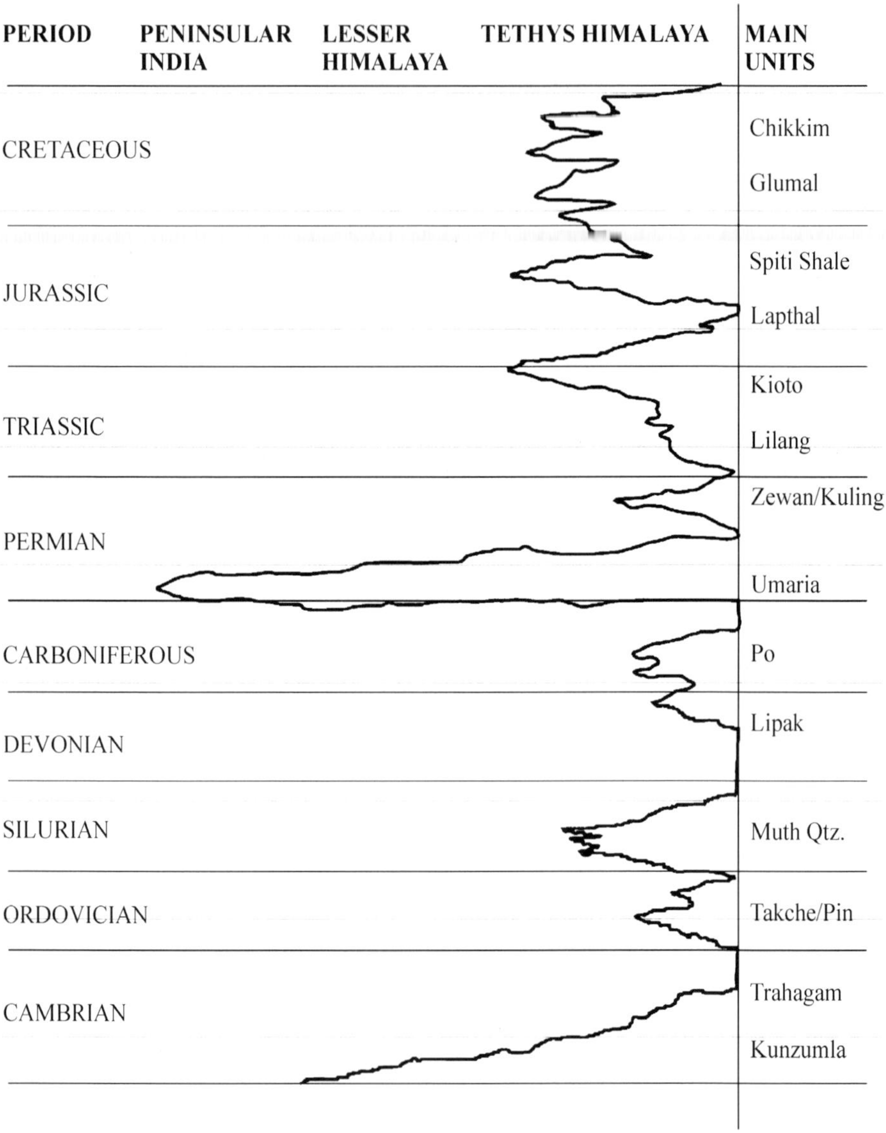

Fig. 6.1 Transgressive and regressive cycles in Palaeo-Mesozoic sequence of India [Redrawn and revised after Shah (1991)]

In Higher Himalaya, some fresh water fauna and flora is also known in Paleozoic times, indicating a localised fresh water deposition which took place during regressive phases in Tethys. Examples of this are the Carboniferous flora in Kashmir and Spiti–Zanskar (lower part of Po Formation) and Early Permian flora and fauna in Kashmir in what are known as *Gangamopteris* beds. However, these are restricted in nature and represent a short span of a probable fresh water or coastal deposition.

6.2 PALEOZOIC MARINE SEDIMENTARY CYCLES

6.2.1 Ediacaran–Cambrian cycle

As discussed in Section 5.3.3, the transgressive cycle which commenced in Ediacaran continued into Cambrian. Accordingly, there was continuous marine sedimentation in Higher Himalaya during Ediacaran and Early and Middle Cambrian. In the Lesser Himalaya, this transgressive event resulted in the deposition of Ediacaran rocks (Krol) and Early Cambrian rocks (Lower Tal of Garhwal; Jhelum Group of Pakistan). The regressive cycle started by the end of Early Cambrian, and Lesser Himalayan region did not receive any sediments after that. However, in Higher Himalaya, the sedimentation continued up to the beginning of Late Cambrian.

The Ediacaran–Cambrian transgression affected only the Higher and Lesser Himalaya and had no impact in peninsular India. Following the regression, the greater part of continental shelf margin, including Higher Himalaya, became a region of positive relief in Late Cambrian and Early Ordovician times. This regressive cycle is marked by an angular unconformity in Spiti–Zanskar and disconformities in other areas. There is no record of sedimentation anywhere in Indian subcontinent in late Late Cambrian and early Early Ordovician times.

There are several opinions that the Late Cambrian–Early Ordovician break in Himalaya marks a phase of diastrophism, comparable and coeval to some episodes of Pan-African orogeny which had affected several adjoining regions. This is also supported by the fact that several granites and granitoids of Himalaya indicate a radiometric age dating around 550–500 Ma, which could correspond to this break (See Section 5.1.3). However, this is yet to be proved conclusively and further data in this regard are necessary.

6.2.2 Middle Ordovician–Silurian Cycle

This is relatively a minor transgressive cycle which resulted in sedimentation only in Higher Himalaya. No corresponding sedimentation is known either in Lesser Himalaya or in peninsular India. The facies is essentially of a shallow marine type, except in north-western part of Kashmir, where some deeper water and turbidite conditions are known in Ordovician part of the sequence (Marhaum Formation).

In Spiti–Zanskar, the sedimentary cycle commences with a thick cross-bedded purple quartzite and conglomerate (Thango/Shian Quartzite). The basal contact is marked by an angular unconformity and the beds transgress a partly eroded Cambrian surface. Over these rocks rests a sequence of highly fossiliferous calcareous rocks (Pin Limestone/Takche Formation) which bear Middle and Late

Ordovician brachiopods, trilobites and corals and some Early Silurian fossils as in certain areas (notably Pin valley) the formation extends into Early Silurian. The final culmination of the cycle is marked by a blanket of time-transgressive quartzite (Muth Quartzite), which is universally exposed throughout the Tethys basin of Himalaya.

An identical sequence exists in Kashmir, except that the Late Cambrian/Early Ordovician break here is of the nature of a disconformity and the Ordovician sequence in the north-western part is of a turbidite facies (as mentioned above). In the south-eastern part, however, the facies is similar to that of Takche Formation of Spiti and bears identical fossils. In the south-eastern part the cycle culminates with a thick suite of quartzite (Muth Quartzite), as is the case in Spiti.

In Garhwal–Kumaun–Nepal region, the cycle is represented by a much thicker shallow marine and coastal deposition, which includes evaporite facies at the base (Garbyang Formation) and carbonate facies (Shiala–Variegated Formations) above. The latter bears a fauna comparable to that of Takche Formation of Spiti. Here also, the Muth Quartzite is the uppermost unit of the cycle.

6.2.3 Middle Devonian–Middle Carboniferous Cycle

There is a long period of non-deposition/erosion following the Ordovician–Silurian cycle, which spans considerable part of Devonian Period. Middle and Later parts of Devonian are represented by the overlying Lipak Formation. Marine rocks of Devonian age are poorly represented in higher Himalaya (except probably in the Dolpo region of Nepal). However, in the trans-Himalayan basin of Chitral there is a classical sequence of marine Devonian rocks. This is not a part of Tethyan facies, since, palaeogeographically, the Palaeotethys sea had considerably widened by Devonian times and Chitral region had distanced itself from the margin of the Indian shield. The latter was an area of positive relief during most of the Devonian Period, while the sedimentation in Chitral was taking place in a deeper part of the sea, far away from Indian shelf.

The Middle Devonian to Middle Carboniferous cycle is a minor sedimentary cycle and marks paralic to shallow marine sedimentation (Lipak and Po Formations of Spiti–Zanskar; *Syringothyris* Limestone and *Fenestella* Shale of Kashmir). It is restricted to some parts of Higher Himalaya: notably, eastern part of Kashmir and western part of the Spiti–Zanskar region. In other regions of Higher Himalaya, there is no sedimentation during this period as well.

The cycle in Spiti commences with a phase of deposition of evaporites, followed by shallow marine carbonates. Towards the later part, there is an association of land plants with shallow marine invertebrate fauna in paralic carbonaceous sediments. In Kashmir, the facies is relatively deeper, but the cycle terminates with plant bearing beds.

6.2.4 Permian Cycle

The commencement of this cycle is marked by two events of tectonic significance. On the one hand, the Proto-Gondwanaland got modified into Gondwana supercontinent, comprising the present-day continents of Africa, South America,

Australia, Antarctica and the Indian subcontinent. The diastrophic activity in this supercontinent resulted in the formation of Gondwana basins, which were sites of fresh water sedimentation throughout Permian and most of the Mesozoic. Along the margin of this supercontinent, the Neotethys came into existence as a continuous sea separating the Gondwana landmass from Laurasia. The opening of the Neotethys was due to rifting, which was attended upon by a widespread effusion of volcanic rocks (Panjal Volcanics) in the western part of Himalaya. The event marked the beginning of the crustal extension, followed by the subsidence of the passive cratonic margin. The rifting seems to be more pronounced towards the western part than that in central and eastern Himalaya. The Indian craton appears to have been hinged along the eastern margin and rotating counter-clockwise. Consequently, rifting was maximum in the north-west. The same movement was also responsible for rifting between India and Madagascar.

On the other hand, the South Pole was located on Gondwana supercontinent in Late Carboniferous and that was responsible for widespread glaciation, resulting in deposition of glacial sediments. Moving of Gondwana away from the South Pole in Permian resulted in climatic warming generating a massive transgressive cycle, causing widespread inundation of the Indian craton. This transgression affected not only the Higher Himalaya but also Lesser Himalaya and peninsular India. In the Lesser Himalaya, the transgressive event continued through Early Permian (fossiliferous Permian of Garhwal, Ghensi Formationof Arunachal, etc.); however, in peninsular India, the transgression is restricted to a short time span in Early Permian, covering only parts of Asselian–Sakmerian (Umaria marine bed and equivalents).

One of the interesting aspects of Himalayan Permian is that middle part (comprising Guadalupian) is unrepresented and there are no fossils of that age. The Permian transgressive cycle in Higher Himalaya, therefore, appears to be truncated in the middle part. However, transgression is resumed in the upper part, though it is restricted to higher Himalaya alone and extends into Triassic and Lias. However, detailed biostratigraphic investigations at the Permian Triassic transition indicate a disconformity at this level. This should correspond to a global regressive stage, at the beginning of Triassic. Accordingly, the Triassic–Liassic cycle should be considered as an independent one, though the regressive phase separating the two may cover a minute time span.

6.3 MESOZOIC MARINE SEDIMENTARY CYCLES

6.3.1 Triassic–Liassic Cycle

During Triassic, the rifting which had commenced in Permian continued and resulted in further widening of Tethys. During this time, the Tibetan sub-plate (Lhasa block) got separated from the Indian plate and rapidly started moving away. The global regressive phase, at the beginning of Triassic, had only a marginal effect in Himalaya. The rate of sedimentation was minimal at the Permian–Triassic boundary but picked up fast in Ladinian, with the increase in the rate of subsidence at the cratonic margin. This situation persisted throughout Higher Himalaya. Towards the northern-most margin, close to the Indus–Tsangpo suture, which was the position of the detachment of the Tibetan sub-plate, the rate of subsidence far outstripped the rate of sedimentation, especially in Norian and Rhaetian and probably in Lias, resulting in the deposition of wildflysch (Lamayuru Flysch, also sometimes referred to as Indus Flysch). This wildflysch bears olistoliths, mainly derived from the Ladinian and Carnian rocks of the same basin. While the subsidence throughout the northern margin was ongoing, its rate varied in space and time and resulted into lateral and vertical facies variation in Kashmir, Spiti–Zanskar, Kumaun and Nepal.

In Spiti–Zanskar, the 'Quartzite Series' towards the top of Triassic, marks a drop in the sea level and rejuvenation of the Indian craton. This process eventually brought the closure of the sedimentary cycle in Liassic, especially towards the western side. This happens to be the last marine cycle in Kashmir, and the Kashmir region seems to have emerged and become a part of the rejuvenated craton after the end of this cycle.

The Triassic–Liassic cycle did not extend either in the Lesser Himalaya or peninsular India, except to the margin of the former, in the north-west (Salt Range area).

6.3.2 Jurassic–Cretaceous Cycle

The pattern of marine transgressions in Jurassic and Cretaceous Periods is different from the earlier ones: in that the transgressions in Himalaya do not exactly correspond with those in peninsular India. This is because the peninsular transgressions were now taking place, not from the Himalayan side but from the newly opened western sea. They followed, more or less, the global pattern, whereas the Himalayan transgressions were governed by the differential subsidence of the cratonic margin, which could be a result of the northwards movement of the Indian plate.

The Himalayan Tethys behaved as a passive continental margin of the Indian plate from Permian onwards; however, further north, along the edge of the craton and beyond, an arc-trench system started evolving in Jurassic and reached its culmination in Cretaceous times. Thus, along the Indus–Tsangpo suture region, the facies of rocks in Jurassic–Cretaceous times is considerably different from that of the Tethys Himalaya and comprises wildflysch rocks (upper part of Lamayuru flysch, Balcha Dhura flysch) which are associated with basic volcanics (Dras Volcanics) towards the end of Cretaceous.

In the Tethyan facies of the Higher Himalaya, there is a break between lower part of Lower Jurassic (Kioto Limestone, Lapthal Formation etc.) and Middle Jurassic sediments (Ferruginous Oolite or Salcacutus beds or Lumachelle Formation). The latter have a limited vertical extension and are succeeded by another break covering major part of the Middle Jurassic. The next transgressive phase comprising black shale (Spiti Shale), representing the Late Jurassic sequence, was due to deposition on an undisturbed outer shelf. The facies, however, changes upwards in Early Cretaceous (Berriasian–Albian) with the coarsening of the sediments. This indicates the progradation of clastic detritus brought in by a deltaic system and described as proximal flysch (Giumal Sandstone). There is a vertical and lateral facies variation, with the bathyal facies (Chikkim Formation, Fotu La Formation) deposited all along the Higher Himalaya from Zanskar to Nepal, where the rate of sedimentation is low with a limited terrigenous supply. Towards the upper part (Santonian–Campanian) pelagic conditions prevail, and all along, the northern margin radiolarian cherts and foraminiferal oozes are associated with it.

In peninsular India, the transgressive phases in Jurassic and Cretaceous are more in consonance with the global eustatic changes and, therefore, do not coincide with those of the Himalayan region. A series of transgressive-regressive cycles are seen in west and south-east of Indian peninsula, producing highly fossiliferous coastal deposits in Kutch, Rajasthan, Baluchistan, Narmada valley and the eastern coast of south India. The main transgressive events in western India are in Bathonian (Patcham), Callovian (Chari), Kimmeridgian (Katrol), Tithonian-Hautrivian (Umia), Aptian (Ukra–Bhuj) and Turonian–Santonian (Bagh). Along the eastern coast of south India, the transgressive phases are those of Cenomanian–Turonian (Utatur), Santonian (Trichinopoly) and Maastrichtian extending into Danian (Ariyalur–Ninyur). Two main transgressive events in early Cretaceous (Valanginian–Barremian) and Late Cretaceous (Santonian–Maastrichtian) resulted in deposition in parts of Baluchistan.

The variation in the timing of the transgressive cycles of Himalaya and peninsular India, and the difference in facies appear to be the main reasons for the distinctness of the fauna, notably ammonites, in the two regions. This has led to ambiguity about the correlation of the two, as also the palaeogeographical interpretations suggesting absence of sea channels, which have been proposed to explain this distinctness. However, in the light of northerly movement of the Indian plate in Jurassic and Cretaceous, it is evident that the transgressions along the Himalayan margin are governed by local tectonics, and the stratigraphic levels of sedimentation are not identical in the two regions, accounting for the variation in fauna.

6.4 CENOZOIC SEDIMENTARY CYCLES

6.4.1 Early Cenozoic Sedimentation

The commencement of the Cenozoic Era was marked by the collision of the Indian plate with the Asiatic plate, resulting in the closure of the Tethys and termination of the shelf sedimentation of Paleozoic and Mesozoic. The initial collision was in the north-western part, and this was followed by the counter-clockwise rotation of the Indian plate. The central region of Himalaya was uplifted and epicontinental marine basins were formed in the eastern and western area, to the south of the newly emerging orogen. These seas, extending from Baluchistan to Garhwal in the west and from Assam to Sikkim in the east, were the basins of shallow marine sedimentation towards the early part of Cenozoic. Moreover, in the north, an epicontinental sea across the line of collision was the site of Eocene marine sedimentation in Ladakh and parts of Tibet. The marine sedimentation was also widespread along the western coast of peninsular India, overlying the Deccan Volcanics.

Following the further uplift of Himalaya, the shallow marine sedimentation yielded place to the sedimentation of the paralic type during Oligocene and early part of Miocene (Murree, Dagshai) in the central region, marine sedimentation, however, continued in the western (Sind–Baluchistan) and eastern (Assam) regions. However, by the end of Miocene, the marine sedimentation was replaced by fresh water sedimentation throughout Himalayan frontal basin.

6.4.2 Late Cenozoic Sedimentation

In the Himalayan region, in Middle Miocene there was a major phase of uplift, which provided the initial relief to the mountain chain. The uplift was responsible for the end of marine or estuarine sedimentation throughout the Himalayan foothill zone, except for the extreme western region in Baluchistan (Manchhar Group) and extreme east in Burma, now known as Mayanmar (Irrawaddy Group). The fresh water sedimentation took place in a foredeep in front of the newly emerged orogen, with the latter supplying the sediment. This was the beginning of the deposition of an enormous group of molasse sediments (the Siwalik Group). This sedimentation continued through Pliocene up to Pleistocene. During deposition, there were other phases of Himalayan orogeny which resulted in uplift and folding of the older units of this Group. The last major phase of Himalayan orogeny in Pleistocene resulted in the final uplift of this Group, which became a part of Himalayan foothills (the Siwalik mountain range). However, the site of deposition of sediments, derived from the mountain chain, shifted southwards on the peninsular margin, which remains a site of deposition throughout Holocene (Indus–Ganga basin).

Apart from the sedimentation in the foredeep, a few intra-montane basins in Himalaya were also the sites of Late Cenozoic sedimentation. The important ones are the Karewa basin in Kashmir and the Kathmandu basin in Nepal. The sedimentation in these basins extended from Late Pliocene through Pleistocene into Holocene.

In peninsular India, Late Cenozoic deposits of fresh water to estuarine type are seen in south India (Rajamundary Sandstone, Cuddalore Sandstone, Warkalli Formation etc.). Most of this deposition took place either on the Precambrian rocks or on the coastal marine deposits of Mesozoic or Early Cenozoic.

Probably the only region in the Indian subcontinent where marine sedimentation prevailed for most of the Cenozoic, extending upto Pleistocene, is in the Andaman-Nicobar Islands. Elsewhere there is no marine sedimentation beyond Miocene. However, in the offshore and some on-shore oil fields of western India, the younger marine rocks have been recorded from the boreholes but no surface outcrops of these rocks exist.

■■■■

7 Ediacaran and Cambrian Sedimentary Succession

7.1 STRATIGRAPHIC SETTING

Ediacaran transgressive cycle started towards the beginning of Ediacaran and affected the northern margin of Peninsular India. Accordingly, marine sediments, which extended in age from Ediacaran to the Early and Middle Cambrian, were deposited. Due to Himalayan structural deformation, these rocks are now exposed in isolated sections, both in Lesser and Higher Himalaya. In Lesser Himalaya, they rest on the Late Proterozoic rocks in Salt Range, Pakistan (Saline 'Series') and Garhwal region (Krol Formation). In Higher Himalaya, they lie in a continuous sequence over the older rocks in Kashmir, Spiti–Zanskar and Kumaun region. There is reason to believe that the basin was a continuous one and the present isolation is purely a structural phenomenon. The transgression extended through the northern part of Indian shield. The Late Ediacaran rocks were deposited throughout this region. The regression started in Early Cambrian and continued through Middle Cambrian, culminating at the start of the Late Cambrian. Accordingly, only Lower Cambrian (and probably a little of Middle Cambrian) and Ediacaran rocks are deposited in Lesser Himalaya, whereas Middle Cambrian and a small part of Upper Cambrian are deposited in Higher Himalaya, in addition to Lower Cambrian and Upper Ediacaran.

7.2 LOWER CAMBRIAN SEQUENCE OF LESSER HIMALAYA

7.2.1 Garhwal

Ediacaran–Cambrian transgressive sequence in Garhwal, as mentioned earlier, comprises Krol–Tal succession. The Upper Krol Formation contains stromatolites viz. *Conophyton, Aldanella, Baicalia, Stratifera, Irregularia,* oncolites, calcareous algae viz. *Renalcis*, *Epiphyton, Olecmia,* acritarchs, in addition to soft-bodied Ediacaran fossils viz. *Pteridinium cf. P. simplex, Charniodiscus sp. Zolotytria sp. Beltanelliformis cf. brunsae, Tirasiana sp., Medusinites sp., Beltanell sp., Kimberella sp., Conomedusites sp., Cyclomedusa sp., Sekwia sp.* and some trace fossils. This confirms the Ediacaran age for these rocks. The Krol Formation grades into overlying Tal Formation (sometimes designated as Tal Group), comprising a thick pile of argillo-arenaceous sequence with chert and phosphorite. The Tal Formation is classified as shown in Table 7.1.

Table 7.1 Cambrian Sequence in Garhwal

CAMBRIAN (LOWER)	UPPER TAL	Upper Quartzite Member
		Limestone Member
		Arkosic Sandstone
		Shale Member
		Lower Quartzite Member
	LOWER TAL	Calcareous Member
		Arenaceous Member
		Argillaceous Member
		Chert (Phosphorite) Member
PRECAMBRIAN (EDIACARAN)	KROL E	Siltstone, shale-limestone interbeds

The thickness of Tal Formation varies and may reach a maximum of over 3000 m. Outcrops of Mussoorie and Nigali Dhar synclines are some of the thickest and best exposed. The rocks are exposed in other areas of Garhwal and Kumaun (Uttarakhand) as well, notably along Ganga valley north of Rishikesh. The Chert (Phosphorite) Member at the base is a horizon marker and serves as a good time rock unit.

The Chert Member contains well preserved skeletal fossils comprising what is known as small shelly fauna. These skeletal remains are denticular structures of doubtful affinities which are widely distributed throughout the globe at the commencement of Cambrian. The taxa reported from Mussoorie section include *Protohertzina anabarica, Anabarites trisulcatus, Tiksitheca korobovi, Spirellus columnorous, Maldeotaia bundalica* and *Circotheca* sp. On the basis of this fauna, these rocks can be correlated to Meischucunian Zone I of China.The rocks also contain calcareous algae: namely, *Obruchevella* and stromatolites known from Tommotian stage of Russia. In addition to these, the Chert Member and the overlying argillaceous member contains an assemblage of acritarchs.

Arenaceous member contains a rich assemblage of trace fossils, the more important ones include *Astropolichnus (Astropolithon)* sp., *Bergaueria* sp., *Cruziana pormensis, Diplichnites* sp., *Monomorphichnus lineatus, Neonereites* and *Phycodes palmatum.* Some body fossils of trilobites are known from the upper part of arenaceous member, notably from Nigali Dhar syncline. Calcareous member contains brachiopods and microgastropods. It is the shale member which bears the characteristic redlichid trilobites in addition to a wide variety of brachiopods, equivalent to Kussak Shale of Salt Range, on the basis of which middle Early Cambrian age can be assigned to it. The topmost quartzite member is extremely thick and marks the culmination of the Cambrian transgressive event in Lesser Himalaya.

7.3 CAMBRIAN SEQUENCE OF HIGHER HIMALAYA

7.3.1 Distribution

As mentioned earlier, the Paleozoic–Mesozoic rocks in Higher Himalaya constitute what is known as the 'Tethyan facies'. In fact, the term is a misnomer because the sequence is not monofacial and comprises various kinds of rocks depending on the depth of sedimentation, time of deposition and nature of the basin. The only difference from Lesser Himalaya is that the deposition is in the outer continental shelf and, therefore, deeper. The transgressive phases are also longer. The Ediacaran–Cambrian transgressive cycle in Higher Himalaya extends up to the beginning of Late Cambrian, whereas it culminates at the beginning of Middle Cambrian in Lesser Himalaya (see Section 6.2.1). The facies being deeper comprises essentially of argillaceous rocks. Since the carbonates are not present, at least towards the lower part of the sequence, the small shelly fauna marking the beginning of Phanerozoic is not reported anywhere in Higher Himalaya.

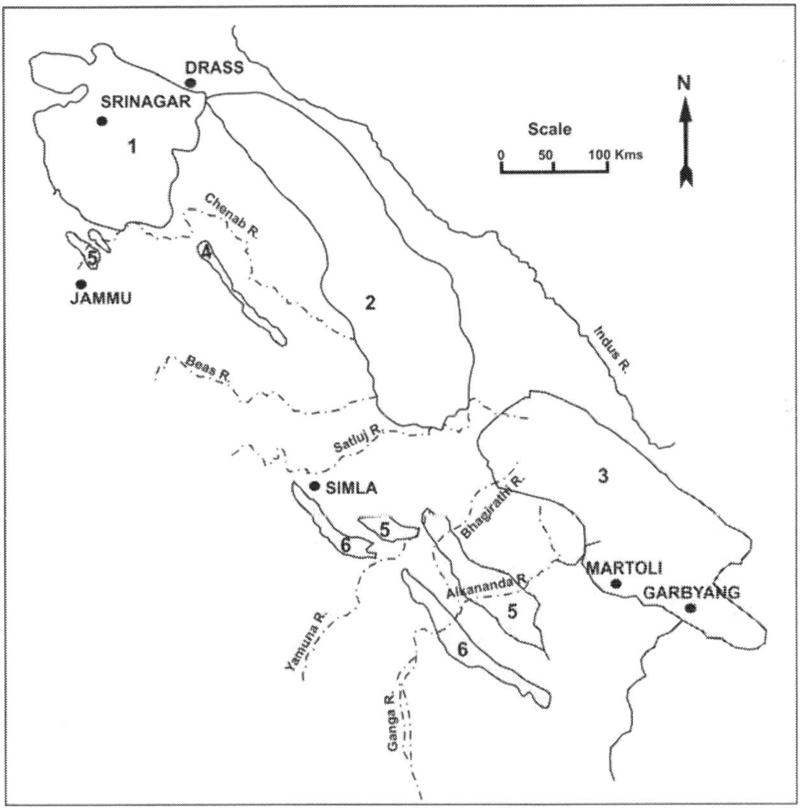

Fig. 7.1 Late Proterozoic, Palaeo-Mesozoic basins of westerm Himalaya.
Tethyan basins: 1. Kashmir basin 2. Spiti–Zanskar basin 3. Garhwal-Kumaun basin.
Lesser Himalayan basins (Proterozoic): 4. Bhadarwah-Chamba basin
5. Shali–Deoban–Tejam–Sirban limestone basin 6. Krol basin

Ediacaran–Cambrian succession in Higher Himalaya occurs as a continuous sequence. In the absence of well-marked fossiliferous horizons of Meischcunian/ Tommotian age, the precise boundary between Precambrian and Cambrian can be only tentatively identified on the basis of appearance of *Treptichnus pedum* in Kashmir. The sequence is well exposed in Kashmir, especially towards the north-west (Kupwara District), Spiti, Zanskar and Kumaun extending into western Nepal. Table 7.2 gives the general distribution and frequently used nomenclature (For basinal distribution in lesser Himalaya refer to Fig. 7.1).

Table 7.2 Distribution of Ediacaran-Cambrian Rocks in Higher Himalaya

PERIOD	EPOCH	KASHMIR		SPITI	ZANSKAR	KUMAUN
		North-west	South- east			
Cambrian	**Late**	TRAHA-GAM	KARI-HUL	KUNZOM LA	KUR-GIAKH	LOWER GARB-YANG
	Middle	NUTUNUS				
	Early	LOLAB			KARSHA	RALAM
Ediacaran		MACHHAL		BATAL	PHE	MARTOLI

7.3.2 Kashmir

Earliest work by Wadia in north-west Kashmir established the occurrence of Cambrian rocks resting conformably on Precambrian, the latter being named by him as Dogra Slate. Since then, more detailed work has been carried out. Cambrian rocks are widely distributed in north-western part of the Kashmir valley, covering major part of Kupwara district. While bulk of the outcrop is concealed by Quaternary lake deposits (Karewa Formation), well-exposed sections are available in both limbs of Marhaum syncline and many other areas in the Lolab valley. The oldest Machhal Formation (equivalent to Dogra Slate) is undoubtedly Late Precambrian in age. The Precambrian–Cambrian boundary lies somewhere within Lolab Formation. Lolab Formation is over 1,700 metres in thickness. Bulk of this is unfossiliferous but there are two distinct trace fossil zones and one trilobite zone, bearing *Redlichia* in addition to brachiopod shells. Nutunus and Trahagam Formations are rich in trilobites and brachiopods of Middle and early Late Cambrian age. As against this in south-east Kashmir (Liddar valley) fossils are extremely scanty where fragmentary trilobites of Middle Cambrian age are found in Karihul Formation at Hapatnar and near Daksum, below Simthan pass. The sequence along with the fossil content is shown in Table 7.3.

Table 7.3 Cambrian Sequence in Kashmir

PERIOD	EPOCH	LITHO-STRATIGRAPHIC CLASSIFICATION	LITHOLOGY	BIOSTRATIGRAPHIC CLASSIFICATION	FAUNA
CAMBRIAN	LATE	TRAHAGAM FORMATION	Bluish and brown crystalline limestone interbedded with greenish slaty shale	*Blackwelderia* Zone	*Blackwelderia* sp. *Cyclolorenzella* sp. *Monkaspis* cf. *serrata*
	MIDDLE	NUTUNUS FORMATION	Greenish and brownish shale interbedded with mudstone and micaceous sandstone	*Damesella* Zone	*Damesella shergoldi*
				Shahaspis Zone	*Shahaspis himalayensis*
				Tonkinella-Bailiella Zone	*Tonkinella breviceps, Bailiella lantenoisi, Hundwarella memor, Baltagnostus* cf. *rakuroensis*
	EARLY	LOLAB FORMATION	Grey to dark brown sandy shale with bands of orthoquartzite	*Redlichia* Zone	*Redlichia takooensis*
				Rusophycus-Cruziana Zone	*Rusophycus* sp. *Cruziana* sp., *Monomorphichnus* sp., *Kupwaria fusiformis, Diplichnites* sp.
				Treptichnus Zone	*Treptichnus pedum, Phycodes palmatum, Planolites* sp

7.3.3 Spiti

Hayden in the beginning of twentieth century discovered the Cambrian rocks in Spiti which he called the Parahio 'Series' since the rocks were well exposed along Parahio (actually pronounced Barachu) river. Parahio Formation constituted the topmost unit of a thick slate formation which was earlier named as Haimanta by Griesbach. The Ediacaran–Cambrian rocks are exposed all along the northern slope of the Great Himalayan Range bordering the south-western side of the Spiti valley. They have been since extensively studied and divided into two formations viz. Batal Formation and Kunzom La Formation, the former unconformably resting on Rohtang Gneissic Complex. The Precambrian–Cambrian boundary should lie somewhere close to the boundary between the two formations, probably within the Kunzom La Formation.

The lower part of Kunzom La Formation, the Debsa Khad Member is 2,100–3,290 metres thick. It bears two distinct trace fossil zones comparable to those known from Kashmir (see Sec. 7.3.2). The upper Parahio Member bears a profuse trilobite population in addition to brachiopods preserved in fine-grained dark grey shale bands separating the coarser silty beds. The fauna ranges in age from Early to Middle Cambrian, since no distinctly Late Cambrian forms have been identified.

Table 7.4 Cambrian Sequence of Spiti

PERIOD / SYSTEM	EPOCH	LITHO-STRATIGRAPHIC CLASSIFICATION	LITHOLOGY	BIO-STRATIGRAPHIC CLASSIFICATION	FAUNA
CAMBRIAN	LATE	KUNZOM LA FORMATION			
	MIDDLE		PARAHIO MEMBER	*Kunmingaspis* Zone	*Kunmingaspis straycheyi, Xingrenaspis maopoensis,*
	MIDDLE		PARAHIO MEMBER: Micaceous sandstone, siltstone intercalated with dark grey shale bands	*Hundwarella* Zone	*Hundwarella memor, Opsidiscus haimantensis, Oryctocephalus salteri*
	MIDDLE		PARAHIO MEMBER	*Pagetia-Oryctocephalus* Zone	*Pagetia significans, Oryctocephalus indicus*
	EARLY		DEBSA KHAD MEMBER	? *Redlichia* Zone	*Redlichia noetlingi*
	EARLY		DEBSA KHAD MEMBER: Thin bedded quartzite with subordinate grey shale	*Rusophycus-Cruziana* Zone	*Rusophycus cf. didymus, Cruziana pormensis, Kupwaria fusiformis, Monomorphichnus lineatus*
	EARLY		DEBSA KHAD MEMBER	*Treptichnus* Zone	*Treptichnus pedum, Phycodes antecedens, P. cf. palmatum, Planolites corrugatus*
EDIACARAN		BATAL FORMATION	Dark grey argillite and greenish grey quartzite		

Unlike that of Kashmir, the biostratigraphy in Spiti has not been amenable to formal zonation. This is because the fauna is widely dispersed throughout the thickness of the fossiliferous sequence and no distinct assemblages can be identified. The zones indicated in Table 7.4 are overlapping and tentative. The only distinct Early Cambrian form *Redlichia* is known from a fallen block along a cliff in Pin valley. All other fossils show Middle Cambrian affinities. The culmination of Parahio Member is marked by a distinct angular unconformity which separates it from overlying Ordovician rocks.

7.3.4 Zanskar

The Ediacaran–Cambrian sequence in Zanskar valley in Ladakh can be subdivided into intertidal Phe Formation, the dolomitic shallow marine Karsha Formation and the silt dominated deeper water Kurgiakh Formation bearing turbidites near the top. The Precambrian–Cambrian boundary would lie within the Phe Formation. While the lithostratigraphy of the sequence has been well established, systematic biostratigraphic data is scanty owing to the difficulty of the terrain. The fauna reported comprises spot samples only, collected from the top unit of Karsha Formation and the basal unit of the Kurgiakh Formation (Surichun Member). This comprises trilobites associated with brachiopods which show late Middle and early Late Cambrian affinities. The more important taxa include the agnostids *Diplagnostus planicauda, Hypagnostus correctus, Lejopyge armata, Gonioagnostus aculeatus, G. spiniger, Linguagnostus* cf. *tricuspis, Clavagnostus* cf. *repandus* and the polymerid trilobites *Fuchouia* sp. *Schmalenseeia amphioneura, Damesops* sp. The fauna is considerably distinct from that of Kashmir and Spiti since it bears a preponderance of agnostids which are rare in the latter two areas. This fauna is essentially of the slope facies and indicates a deeper water environment as compared to that of Spiti. The fauna also appears to have been collected at a higher stratigraphic level which could account for the difference.

Karsha Formation towards the lower part bears a rich assemblage of trace fossils comparable to those of Spiti and Kashmir. As in latter two areas, two well marked assemblages can be distinctly identified. The lower assemblage bears a profusion of burrow fillings mainly *Planolites* with occasional preservation of *Phycodes palmatum.* The upper assemblage, in addition to the burrow fillings also bears trails and traces of arthropod origin such as species of *Rusophycus, Cruziana, Monomorphichnus* and *Diplichnites*.

7.3.5 Kumaun

In Kumaun, the Ediacaran–Cambrian rocks rest on Crystallines of higher Himalaya (named as the Vaikrita Group). The contact is usually a tectonic one, especially towards the western side, due to which the bulk of the older part of the sequence appears to be truncated. As in the rest of the Higher Himalaya, the lower part of the sequence comprises dark grey shale and slate, low grade carbonaceous phyllite with bands of quartzite, named as Martoli Formation. This is overlain by a unit of alternating purple quartzite–conglomerate sequence, the Ralam Formation, which in turn is overlain by Garbyang Formation. The latter has an enormous thickness ranging from 800 metres to well over 3,000 metres. It comprises alternating bands of calcareous shale and dolomite towards the lower part and laminated quartzite, graded sandstone and green shale towards the upper part. The three formations considerably vary in individual thickness from east to west and apparently have a time transgressive relationship, the entire sequence having been deposited in the inner shallow shelf and partly in the intertidal environment.

Compared to other Higher Himalayan sections, the fossil record in Kumaun is poor. It is only in recent years that the redlichid trilobite horizon was discovered in Martoli in the Milam–Topidunga section. Prior to that, only trace fossils were known from the lower part of Garbyang Formation. The Precambrian–Cambrian boundary should lie somewhere within the Martoli Formation. The upper limit of the Cambrian in this region is doubtful in the absence of any distinctive fossils. It could lie within Garbyang, probably along the disconformity between lower and upper Garbyang, though some people believe that the boundary should be below Ralam, rendering the entire Garbyang as Ordovician, in addition to overlying Shiala Formation which is known to be undoubtedly Middle Ordovician. This opinion seems to be rather far-fetched.

7.4 PRECAMBRIAN–CAMBRIAN BOUNDARY

7.4.1 Boundary on the Global Scale

The identification of Precambrian–Cambrian (Pc–C) boundary in India can be based only on the criteria of its recognition on a global scale. After a debate extended for several years the Sub Commission of Stratigraphy of IUGS accepted Fortune Head section of Burin Peninsula, Newfoundland (Canada) as Global Stratotype Section and Point (GSSP) for this boundary. The other contenders for the GSSP were Meishcun section in China and Aldan river section in Siberia. The problem with the fixation of the criteria was that all these three sections represent different facies of deposition and, therefore, the fauna in each of them is different. The Fortune Head Section is rich in trace fossils while Meishcun section and Aldan river section are rich in small shelly fauna and archaeocyathids, respectively.This was the main hurdle in fixing a GSSP, since the known organisms have strong facies control at the Pc–C level. One of the reasons for the choice of trace fossils as significant for the fixation of boundary was that trace fossils were widespread in all facies and were not facies restricted like small shelly fauna and archaeocyathids. Accordingly, they could be more useful in global correlation.

The history of Pc–C boundary fixation has passed through many vicissitudes during last few decades. At one time, it was believed that fossilisation started only from Cambrian onwards. In fact, that accounts for the term Phanerozoic (visible organisms). Advanced scientific techniques have made it possible to observe and study fossils way below in the Precambrian, though it is true that skeletelisation on a large scale started only in Cambrian. For many decades, the boundary was taken to be marked by the first appearance of trilobites. While appearance of trilobites is a landmark and still a basic guideline for boundary fixation in that trilobite-bearing rocks are positively Cambrian, their first appearance is no longer considered as marking the boundary. This is because many skeletal organisms appeared before them and if skeletalisation is the criterion, then trilobites are not the pioneers and, therefore, do not deserve to be the markers of an Erathem boundary. However, one of the definite and established premises about the boundary is that it must lie above the Ediacara fauna and below the first appearance of trilobites.

The various fossil groups which are significant for the boundary fixation are small shelly fauna (SSF) and archaeocyathids for their first appearance, trace fossils for their continuous evolution and vendotaenides and soft bodied *Charnia* fauna and medusoids for their disappearance, though the last could not be synchronous. In addition to this, there could be a geochemical basis for boundary identification.

Small shelly fauna constitutes a group of small tubular and horn-like structures of phosphatic composition whose affinities to the organisms to which they belong are uncertain. They are probably denticular structures of some predators that appeared suddenly at the dawn of Cambrian and were probably responsible for the decline of vendotaenides, a group that flourished after Ediacara fauna. The appearance of SSF being the beginning of skeletalisation event could mark the commencement of Cambrian. However, the use of SSF as a biostratigraphic criterion for identification of the Pc–C boundary has not been satisfactory mainly owing to their high degree of provinciality and a strong facies preference. They are palaeogeographically restricted to only three belts in present-day Asia. Biostratigraphically, their best preservation is in Meishcun section of China. They are also restricted only to carbonate rocks and this lithological preference hampers correlation in such regions where carbonates are not deposited at this level. In addition, the appearance of SSF seems to be unrelated to the evolutionary sequence of other biota which existed before or appeared after them.

Archaeocyathids are a group of reef builders which appeared suddenly and, therefore, have a potential of being significant boundary markers. Being reef builders they thrived only in shallow waters on the flanks of carbonate belts. This limited distribution and strong facies preference adversely affects their utility for boundary fixation. At the boundary level, they are best known from the Aldan section of Siberia and hardly from anywhere else.

Ichnofossils or trace fossils, as the name suggests, are the traces (trails, burrows, resting excavations etc.) of soft bodied organisms. Accordingly, they are found everywhere i.e. in Precambrian as well as in Cambrian and post-Cambrian rocks. Their preservation does not have a marked facies preference, though better preservation is available in alternating sandstone-shale sequence. While there is a variety of various ichnotaxa and an abundance in population, bulk of them are long ranging. It is only a few which could be characterised as horizon markers for Pc–C boundary. Their greatest advantage, apart from the less marked facies control in Lower Cambrian, is that they represent a continuous evolutionary sequence of organisms and, therefore, constitute a best tool for evolving a bio-event stratigraphy. During last decade or so the significance of trace fossils: notably, the first appearance of *Treptichnus pedum* (originally known as *Phycodes pedum* and sometimes referred to as *Tricophycus pedum*) and related forms, for the fixation of Pc–C boundary has been widely appreciated, which accounts for recognition of Fortune Head Section, Burin Peninsula in Newfoundland as GSSP.

The geochemical factors for Pc–C boundary fixation are still not fully established. The geochemical changes, which may be related to any significant events, comprise mainly, decrease in dolomite accumulation, widespread appearance of red biogenic limestone and a global phase of phosphogenesis. Additionaly, an important factor which has widespread ramifications at other Erathem boundary levels, is the heavy excursion of $\delta^{13}C$ in terminal Precambrian dolomites. The succeeding phosphorites have high $\delta^{13}C$ values. This could be a result of explosive expansion of heterotrophic organisms leading to increase in carbon respiration resulting in the liberation of ^{12}C. The rapid expansion of these organisms caused the generation of phosphorus.

7.4.2 Precambrian–Cambrian boundary in India

From the global perspective given above it is clear that two of the three faunal assemblages which are significant for the fixation of boundary viz. SSF and archaeocyathids are facies controlled. Of these archaeocyathids are not known anywhere in India. However, the SSF has been reported from Krol–Tal junction in Mussoorie. The trace fossils are found in all sections in Higher Himalaya as well as Lesser Himalaya and are probably more suitable for the identification of the boundary.

The discovery of Ediacaran fauna from the topmost Krol unit in Nainital clearly indicates that the Pc–C boundary must be above Krol, possibly at the base of Tal Formation. The pre-trilobite fauna in Tal Formation comprises mainly the small shelly fauna from chert–phosphorite member. This is comparable to the Meishcun section of China as far as the facies is concerned. However, Meishcun section represents a monofacial sequence where biostratigraphic ranges of various taxa can be seen. In Mussoorie, the bulk of SSF is restricted to chert–phosphorite member and this restriction is not stratigraphic but facies controlled, as the underlying and overlying rocks are different. The correlation with Chinese section is only broad and not precise. Despite that, it can be safely inferred that the boundary is close to the base of Tal and more or less coincides with the Krol–Tal junction.

As against Lesser Himalaya, Ediacaran–Cambrian sequence in Higher Himalaya is by and large monofacial in Kashmir, Spiti and Zanskar. Since there are hardly any carbonates, SSF is not found in these rocks, but a wide range of trace fossils is known in all these sections. In that respect, these sections are comparable to the sequence in Burin Peninsula, Newfoundland. The first appearance of *Treptichnus pedum* and associated ichnofauna in these rocks marks the commencement of Cambrian. The *Treptichnus pedum* Zone bears a wide variety of grazing and burrowing trails which are considerably different from simple burrows preserved in underlying rocks. This indicates an expansion and diversification event at this level and should be the main criterion in the identification of the boundary. Regarding ichnostratigraphy, the Higher Himalayan sections compare favourably with those of Burin Peninsula and, therefore, afford a global correlation of the boundary.

7.5 THE CULMINATION OF THE EDIACARAN-CAMBRIAN CYCLE

The Ediacaran–Cambrian transgression was at its maximum in Early Cambrian times when large areas in Lesser Himalaya were covered by the sea in addition to those of Higher Himalaya. The rocks deposited in Lesser Himalaya are those of shallow inner shelf while those in Higher Himalaya represent outer shelf or even slope facies. By Middle Cambrian, the sea had already receded from Lesser Himalaya but the deposition in Higher Himalaya was continuing. A small thickness of Upper Cambrian rocks was deposited in Kashmir and probably in Zanskar but bulk of Upper Cambrian is not known anywhere in India. In addition, the boundary between Middle Cambrian and overlying Middle Ordovician in Spiti is marked by a distinct angular unconformity. The conglomerate overlying this unconformity bears pebbles of older granite and quartzite of Batal and Kunzom La Formations. In Kashmir, there is no angular unconformity but a disconformity nevertheless is apparent from the sudden overlap of rocks bearing Middle Ordovician fauna over lower Upper Cambrian rocks. A similar situation is also available in Zanskar. In Kumaun the upper limit of Cambrian is doubtful. According to one opinion, it is marked by Ralam Conglomerate while according to others it is marked by the unconformity between Lower and Upper Garbyang. In either case an unconformity marks the boundary. This break which eliminates Upper Cambrian and Lower Ordovician rocks throughout Himalaya is a crucial one. It coincides with the pan-African orogeny and there is a general opinion that this orogeny had an impact in India as well. This is also supported by radiometric ages of many higher Himalayan granitoids and migmatites which suggest that a significant tectonic event occurred around 500 Ma.

■■■■

8 Post-Cambrian Paleozoic Stratigraphy

8.1 POST-CAMBRIAN CYCLES IN PALEOZOIC SEQUENCE

Throughout the Indian sub-continent, there is no record of sedimentation in late Late Cambrian and Early Ordovician. As mentioned earlier (Sec. 7.5) this could be a result of an orogenic event. The well-marked angular unconformity in Spiti followed by alluvial fan conglomerates and sandstone, also known from Zanskar (variously named as Shian Formation, Thango Formation, Thaple Formation) bearing clasts derived from the erosion of newly formed orogenic belt, are strong evidences in support of this contention. A mild transgressive event extending from Middle Ordovician to Middle Silurian affected Higher Himalaya (see Sec. 6.2.2). After the regressive phase marked by Muth Quartzite, there is no sedimentation in Late Silurian and most of the Devonian. The only undoubted marine Devonian rocks are deposited in the trans-Himalayan region of Chitral (Pakistan). There are reports of late Middle Devonian conodonts in Spiti and in Nepal. (see Sec. 6.2.3).

The next transgression, affecting only some parts of Higher Himalaya, took place towards late Middle Devonian through Early Carboniferous extending into Middle Carboniferous. Late Carboniferous again marked a break which corresponds with the commencement of glaciation in Gondwanaland. However, the transgression which commenced in Early Permian affected not only Himalaya but also peninsular India (see Sec. 6.2.4).

8.2 ORDOVICIAN–MIDDLE SILURIAN CYCLE IN HIGHER HIMALAYA

8.2.1 Kashmir

The Ordovician–Silurian rocks in Kashmir comprise two different facies: one exposed in north-western Kashmir in the core of the Marhaum syncline (Kupwara District) and the other in south-eastern Kashmir in Anantnag District. The rocks deposited in north-western part, the Marhaum Formation, represent outer shelf to slope sedimentation. The lower part of the sequence comprising shale-limestone alternations constitutes shelf deposits. The shale bands in this part are rich in brachiopods: notably *Orthambonites, Dalmanella, Rafinesquina* and *Leptaena.* Towards the upper part, deepening of the basin is indicated by the deposition of turbidites, including graded-bedded greywackes. The sedimentary evolution testifies to strong subsidence with gradually reduced sediment yield. This part of the formation contains only occasional and fragmentary fossils. The fossils are indicative of Middle to Late Ordovician age. There are no rocks younger than these exposed in Marhaum syncline, though a small thickness of Muth Quartzite is exposed in the adjoining Shamsabari syncline.

In south-eastern Kashmir, the Ordocian–Silurian sequence comprises Margan Shale, Gugaldar Formation and Muth Quartzite, representing a transgressive-regressive cycle. These rocks are widely distributed in Lidar valley and upper reaches of Pir Panjal range in Anantnag District. Margan Shale is thickest (over 300 metres) along Margan Pass but its thickness ranges between 30 and 50 metres elsewhere. It passes vertically and laterally into Gugaldar Formation. This variation and nature of sediments indicates a deposition in extremely shallow inner shelf in sub-tidal zone. The lithology comprises calcareous and ferruginous sandy shale. These rocks pass gradually upwards into Muth Quartzite. Margan Shale is rich in fossils, predominantly brachiopods, the essential elements including *Dalmanella basalis.* The fauna shows a range from Middle Ordovician to Early Silurian. The overlying Muth Quartzite is over 800 metres thick in Kashmir. It comprises massive cream-white quartzite which must have been rapidly deposited in a regressive sea. No fossils are known from Muth Quartzite in Kashmir, but on the basis of the fossils from Margan Shale, the age could not extend beyond Silurian.

8.2.2 Spiti

As mentioned earlier (Sec. 8.1), the basal unit of Ordovician in Spiti comprises a quartzite–conglomerate sequence named as Thango Formation (=Shian Formation). This overlies the Cambrian rocks (Kunzom La Formation) with a distinct angular unconformity. It consists of medium to thick bedded, purple, brick red and occasionally grey, current bedded quartzite with intraformational conglomerate. The conglomerate at the base of this Formation contains granite–gneiss pebbles which constitute 2%–3% of the total pebble content. This would indicate uplift in the source area during this period (see Sec. 7.5). The well sorted and cleaner nature of the quartzite and the cross-stratification suggest deposition under high energy, near shore conditions. A part of sequence could even be of aeolian beach origin. The Formation is totally devoid of fossils. Since overlying Formation bears Middle-Late Ordovician fossils, the formation could be of Middle Ordovician age.

Overlying the Thango Formation with a gradational contact is a fossiliferous shale-limestone sequence, the Takche Formation (=Pin Limestone). The lower part of the formation comprises thin bedded purple and green siltstone, variegated shale with bands of grey quartzite. The siltstone shows occasional mudcracks, rain-prints and bioturbation structures while quartzite bears a wide variety of asymmetrical ripple marks and herringbone cross stratification. This would point to a deposition in intertidal conditions. The upper part of the formation shows a wide vertical and lateral variation from quartzite, calcarenite, argillaceous limestone to dolomitised reefal limestone. There is a remarkable decrease in thickness of the lower areneceous-argillaceous part of the formation from north-west (Kunzom La–Takche section) to south-east (Pin–Parahio section) and a corresponding increase in the upper calcareous part. The characteristic part of the formation is the reefal limestone which is a stromatoporoid-algal complex. Both the limestone and shale units bear profuse fossils, mainly brachiopods, though trilobites, bryozoans and algae are also preserved. The more important taxa include the brachiopods, *Dalmanella, Rafinesquina, Leptaena, Dinorthis,* trilobites *Illaenus, Flexicalymene*, bryozoan *Dianulites* and dasycladacean algae, *Coelosphaeridium, Apidium.* These are indicative of Middle to Late Ordovician age.

As in case of Kashmir, the Muth Quartzite overlies the Takche Formation. Lithologically, it is the most striking formation of the entire Spiti region. Its contact with the underlying rocks is marked by thin-bedded quartzite. However, bulk of the formation comprises thick-bedded white and greyish-white quartzite which stands out in relief marking a characteristic topographical feature. The formation represents a regressive part of the cycle comprising beach sands deposited as a blanket on the inner part of the stable shelf. One of the most characteristic features of this formation, as well as the underlying Takche Limestone, is their time-transgressive nature. The upper and lower limits of Muth Quartzite are not the same throughout Spiti region. This is amply clear from the occurrence of characteristic fossil horizon bearing *Pentamerifera* (originally described as *Pentamerus*). This occurs at the base of the formation in Pin valley but occurs towards the top in Kunzom La–Takche section. *Pentamerifera* is indicative of a Late Silurian (Ludlovian) age. The time-transgressive relationship is also seen in underlying Takche Formation where fossil evidence indicates its upper limit as Late Ordovician in Kunzom La–Takche section and probably Early-Middle Silurian in Pin-Parahio section. Nature of deposition of Muth Quartzite indicates that it was rapidly deposited and, despite its thickness, represents a minor time span. There is no evidence of its extension in Devonian.

8.2.3 Zanskar

The Ordovician–Silurian sequence in Zanskar is identical to that of Spiti, except for the fact that the sequence is much more truncated and the equivalent of Takche Formation is absent. The Muth Quartzite is also considerably reduced in thickness. There is a tectonic contact between the Cambrian Kurgiakh Formation and overlying Thango Formation (locally known as Thaple Formation). As in Spiti, the Thango Formation comprises red conglomerates and sandstones and has a thickness of approximately150 metres. It is overlain unconformably and sometimes with a tectonic contact by Muth Quartzite, thereby eliminating the Takche Formation.

8.2.4 Kumaun

In Kumaun, the precise location of the boundary between Cambrian and Ordovician remains unclear. This is because, unlike that of Spiti, the Cambrian trilobites are extremely rare. There is a report of a single *Redlichia* in Martoli Formation (see Sec.

7.3.5). The boundary should lie between Lower and Upper Garbyang, where there is a distinct disconformity and change in the facies. The upper Garbyang comprises cross-bedded calcareous sandstone followed by sandstone-shale alteration. A flat gastropod doubtfully identified as *Ecculiomphalus* is the only fossil reported from this part of the formation.

Garbyang Formation is conformably overlain by Shiala Formation comprising green shale with bands of arenite towards the lower part which, with a gradual increase in the arenaceous component, becomes essentially quartzitic in the upper part. The formation bears frequent bands of biostromal limestone consisting of fragments of crinoids, calcareous algae, corals, bryozoans and brachiopod shells. The formation is rich in fossils notably brachiopods *Rafinesquina, Leptaena, Dalmanella* alongwith some bryozoans and trilobites. The fauna is comparable to that of Takche Formation of Spiti and is indicative of Middle to Late Ordovician age.

Overlying Shiala is one of the most remarkable formations of Kumaun named as the Yong Limestone. This is a reefal limestone complex characterised by a helmet-like disposition discordant with the bedding, the individual helmets being connected with one another by bedded reefal debris. The complex is made of calcareous algae, stromatoporoids, crinoids, corals and bryozoans. The equivalent of this formation in Spiti is the top-most thin reefal band in Takche Formation (see Sec. 8.2.2). However, Yong Limestone is thick and mappable and, therefore, deserving an independent formational status. The age cannot be younger than Late Ordovician.

Between Yong Limestone and Muth Quartzite another Formation named, rather erroneously, as Variegated Formation is widely distributed throughout Kumaun Himalaya. It consists of multicoloured shale alternating with limestone. It has maximum thickness towards east but thins out to the west. It is poorly fossiliferous and is probably time transgressive. The fossils include fragmentary dolerorthids. The Formation could be doubtfully assigned an Early Silurian age.

Muth Quartzite, as in Spiti and Kashmir, is a characteristic Formation in Kumaun, contrasting in colour with the underlying formations. The lower part of the formation comprises chocolate brown quartzite interbedded with bands of dolomitic limestone. This is followed by massive white sugary quartzite. The lower part bears *Pentamerifera* horizon. Accordingly, the age is Silurian as in the case of Spiti. There is no evidence that the formation extends into Devonian.

8.3 DEVONIAN BREAK IN HIMALAYA

Throughout Himalaya Late Silurian and most of Devonian period marks a perceptible break. Except for some possible Devonian fauna reported from Dolpo in Nepal and Middle-Late Devonian conodonts notably species of *Icriodus* from basal units of Lipak Formation in Spiti, no definite Devonian fauna is known anywhere else. The Devonian break also corresponds to a fall in global eustatic towards the early part of the period and in Himalayan region it marks a major period of regression. The Himalayan region had probably emerged during this period and was an area of positive relief. However, no fresh water formations are known either. From trans-Himalayan region of Chitral, a well-developed marine and typically endemic fauna of Devonian is known. However, this region was far away from Palaeotethys and probably represented a deeper part of the sea.

8.4 MIDDLE DEVONIAN-MIDDLE CARBONIFEROUS TRANSGRESSIVE CYCLE IN HIGHER HIMALAYA

8.4.1 Extent of the Cycle

This cycle is by far the least spread cycle in Higher Himalaya in that its extent is limited to some parts of Kashmir, Spiti and Zanskar. It is represented by an extremely shallow marine sequence which is intercepted by a series of diastems. In range, it extends from Givetian in Spiti and Tournaisian in Kashmir to probably Bashkirian, though the upper limit cannot be clearly demarcated. Since the facies represents shallow shelf, it is rich in benthonic fauna. There have been several oscillations during the deposition which are represented by frequent alternations of coastal quartzite and shale bands bearing land flora. The distribution is also erratic and in most sections the sequence is absent. The thickness of the deposits, wherever present, is highly variable. This cycle has affected only certain areas, even within the same basin. The deposition is also marked by a series of diastems and paraconformities as can be seen by rapid facies variations throughout the sequence. The sedimentation on the whole was extremely close to the shore and sometimes in lagoonal conditions.

8.4.2 Kashmir

The Early and Middle Carboniferous rocks in Kashmir are deposited only in south-eastern part of the basin i.e. Anantnag district and parts of Pir Panjal range. They are absent in north-western part where Ordovician rocks are followed directly by rocks of Permian age. Evidently, during Carboniferous times, bulk of Kashmir basin was an area of positive relief and marine transgression was restricted to a small region.

The Lower and Middle Carboniferous rocks in Kashmir comprise what are called[1] *Syringothyris* limestone and *Fenestella* shale. The former comprises bedded and massive limestone alternating with quartzite. The latter consists of shale bands alternating with thick quartzite. In fact, the name shale is a misnomer for this formation as bulkof it comprises quartzite, but since fossils are present only in the shale bands, hence the name. These two rock units are widely distributed throughout Lidar valley in Anantnag district and along Pir Panjal from Margan Pass through Bannihal, Gulabgarh to Budil.

Syringothyris limestone bears the characteristic brachiopod *Syringothyris cuspidata* and a wide variety of other brachiopods—notably, *Linoproductus*—in addition to poorly preserved corals and crinoids. The fossils are ideally preserved in

---

[1] The terms *Syringothyris* limestone and *Fenestella* shale are informal as they do not conform to code of stratigraphic nomenclature. However, because they are firmly entrenched in literature and are mappable and easily identifiable units, it is advisable to retain the names as such.

bedded limestone which constitutes the lower part of the formation. In Lidar valley, the *Syringothyris* limestone is separated from overlying *Fenestella* shale by quartzite which is sometimes referred to as 'passage beds'. Recently poorly preserved fragments of plant fossils referrable to lycopsids have been found in these beds. These have been tentatively identified as *Lepidodendropsis, Lepidosiggilaria* and *Archaeocalamites.*

Fenestella shale, as mentioned earlier, comprises thick massive quartzite separated by thin bands of dark calcareous shale. The shale bands are profusely fossiliferous of which the bryozoan *Fenestella* is the most conspicuous member. Other elements include another bryozoan *Protoretepora*, brachiopods *Spiriferella, Dielasma,* various kinds of productids, notably *Linoproductus, Buxtonia*, bivalve *Modiola* and trilobite *Phillipsia.*

8.4.3 Spiti

In Spiti, as in Kashmir, the Upper Devonian–Middle Carboniferous rocks are not uniformly distributed. While they are present in Losar area in north-west and Po to the east, they are not preserved to the west in Pin and Parahio (Barachu) valleys. The lithology and fauna is also comparable. The lower unit, known as Lipak Formation, comprises thin and medium bedded grey limestone alternating with calcarenite. The bottom-most unit is reported to bear conodont species of genus *Icriodus* indicating a late Middle Devonian (Givetian) age. Middle and upper parts of the formation, however, bear fossils of Early Carboniferous age. Towards the top, this formation represents an evaporite facies containing gypsum and dolomite interbeds indicative of partly dessicating conditions under an arid climate. This is also indicative of a regressive sub-cycle which culminates with the overlying plant bearing bed. Lipak Formation bears *Syringothyris cuspidata, Linoproductus, Buxtonia* and other brachiopods comparable to those found in *Syringothyris* limestone of Kashmir. It also bears a wide variety of conodonts *Bispathodus, Clydagnathus* and *Polygnathus.* The age of the formation spans probably from Middle Devonian to Early Carboniferous.

The upper unit in Spiti, known as Po Formation, comprises laminated shale, black carbonaceous shale full of concretions, alternating with thick quartzite and occasional dolomite bands. Fossils are extremely rare in the formation and only a few shale bands bear profuse fossils, while bulk of the rest is devoid of them. Two horizons of plant fossils, one at Losar to the north-west and the other at Thabo to the east, probably representing two different levels are known. These fossils are preserved in black carbonaceous shale which is the common rock type, especially towards the lower part of the formation. The Losar horizon, which is the older one, bears *Lepidodendropsis schurmani*, while the younger horizon exposed at Thabo bears *Rhacopteris ovata.* The former is indicative of Visean age while the latter shows Middle Carboniferous affinities. The marine fossils, predominantly brachiopods and bryozoans, are profuse in some of the shale bands towards the top of the formation. The fossils include *Fenestella, Protoretepora, Buxtonia, Cleiothyridina, Dielasma* in addition to several other forms. The lithology and flora of Po Formation is indicative of beach-lagoon deposition and shows proximiy to the shoreline towards the lower part. The quartzite towards the upper part is of the nature of shoe-string sands separated by marine shale.

8.4.4 Zanskar

The Lower and Middle Carboniferous rocks in Zanskar are identical with those of Spiti in all respects, except that no plant horizons have been reported from there. The Lipak Formation comprising primarily limestone bears several gypsum bands. Po Formation is mainly quartzite with bands of carbonaceous shale. The fauna of both the formations is identical to that of Spiti.

8.5 LATE CARBONIFEROUS BREAK AND ITS SIGNIFICANCE

Late Carboniferous has been a time of major changes, not only in the Indian sub-continent but in the entire group of continents aggregated together into what is known as the Gondwana supercontinent. The Gondwana supercontinent comprising India, Australia, Madagascar, Africa, South America and Antarctica was at this point of time close to the South Pole and was eventually covered by enormous ice sheets. Consequently, no sedimentation took place during this time and that accounts for a universal break in all parts of this land mass. This was also the period when Palaeotethys expanded from a local epicontinental sea to a continuous oceanic barrier, separating the southern Gondwana supercontinent from the northern Laurentia. The movement of the glaciers scooped huge valleys which later on became the sites for fresh water sedimentation. In India, the highest topographic feature at this time was the Aravalli range. Glaciers from this range moved in all directions when they eventually started melting, resulting in the deposition of tillite towards the beginning of Permian.

Important tectonic changes were taking place during this time along the margin of the Gondwana supercontinent bordering Tethys. This was the time when Alpine region and many other areas of the globe were passing through Hercynian orogenic episode. This episode was responsible for the creation of Tethys. In India, while there was no direct impact of the Hercynian orogeny, the epeirogenic movements resulted into large scale block faulting which continued through Permian and major part of Early Mesozoic. This was the time when the Lhasa block (Tibet) started separating from the Indian plate. The tearing of the block was attended upon by island arc volcanism in north-west (Panjal Volcanics) and north east (Abor Volcanics). This volcanism continued through Permian and extended in a subdued manner in Triassic as well.

8.6 PERMIAN TRANSGRESSIVE CYCLE

8.6.1 Commencement of the Cycle and Peninsular Incursions

Melting of the glaciers towards the beginning of Permian in the entire Gondwana resulted in large scale sea-level changes. Thus, not only the coastal regions to the north bordering the Indian peninsula were inundated, but the sea incursions also extended within the peninsula itself, though only for a short period of time. The Early

Permian transgressive cycle was by far the most widespread cycle which affected the entire Indian Subcontinent. It equally affected other parts of Gondwana supercontinent, and large-scale marine incursions took place in other continents. In western Himalaya, the transgressive event was accompanied by volcanic activity, which commenced with pyroclastic outbursts and volcano-sedimentary deposition associated with glacio-marine sediments. In Peninsular India, the tillite deposits were associated with shallow marine sediments. These sediments, which were first reported from Umaria coalfield and came to be known as 'Umaria marine beds', have since been reported from many other regions in central, eastern and western India, indicating that the incursion was widespread.

While the marine incursion in Peninsular India and even Lesser Himalaya was short-lived, in Higher Himalaya the transgressive event continued through Permian. However, there is no fossil evidence of middle part of Permian (Guadalupian) in Himalaya, though the upper part (Lopingian) is extremely well represented. It is customary to divide the Early Permian marine rocks in two facies viz. marginal marine facies and Tethyan marine facies based supposedly on the depth of the basin, as evidenced from the nature of the fauna and association of Gondwana plant-bearing beds. The two facies do not exactly coincide with the peninsular and extra-peninsular division. The marginal marine facies is known from many areas in peninsular India—notably, Umaria, Manendragarh and Bap—while from the extra-peninsular area, it is known from Salt Range (Pakistan), Arunachal Pradesh, Sikkim, Himachal Pradesh and Kashmir. These marine rocks are associated with Talchir Group, a formation of glacigene character, at least towards the lower part. In almost all these areas the commencement of the marine sequence was characterised by *Eurydesma–Deltopecten* horizon, which, in addition to being indicative of shallow coastal environment, also indicated a cool sea in Early Permian times. The fauna in these localities is not identical, with the dominant forms varying from locality to locality, brachiopods sometimes being the abundant forms and sometimes bivalves. The two famous genera *Conularia* and *Eurydesma* seem to be mutually exclusive, except in the Salt Range. Despite these local peculiarities, sufficiently characteristic species occur at each locality to indicate the contemporaneity. In addition to bivalves *Eurydesma* and *Deltopecten*, scyphozoan *Conularia*, the other dominant faunal elements include brachiopods *Linoproductus, Buxtonia, Dielasma* and *Trigonotreta.* This fauna is identical with transgressive coastal deposits in other parts of the Gondwana superecontinent notably Lyons Group of the Carnarvon basin and Holmwood Shale of the Irwin River basin of W. Australia. The fauna is indicative of Sakmarian age. The peninsular marginal marine sediments are restricted only to this age, though in Higher Himalaya the sedimentation continued constituting the Tethyan facies. The distinction between marginal marine facies and Tethyan marine facies is only with respect to time of deposition and does not seem to have much of a geographical significance since the faunal elements of marginal facies are also present in basal Permian of Tethyan facies. The two facies evidently mark Lower Permian and Upper Permian time spans with Middle Permian being absent. Accordingly, the term marginal facies used here refers to the Early Permian transgressive phase that bears *Eurydesma–Deltopecten* fauna.

8.6.2 Permian Marginal Facies in Peninsular India and Lesser Himalaya

In peninsular India, the Early Permian transgression was widespread and marine deposits associated with Gondwana sediments or otherwise are known from widely separated areas like Umaria and Manendragarh in Madhya Pradesh and Bap in Rajasthan. The beds in Umaria and Manendragarh are 3 and 5 metres thick, respectively, and contain *Linoproductus, Trigonotreta, Protoretepora, Spiriferina* in addition to *Eurydesma* and *Deltopecten*. In the Bap region of Rajasthan, Badhura Formation bears *Conularia* in addition to several bivalve genera.

In Lesser Himalaya, the early Permian marine sediments were deposited in what constitutes the peri-Gondwana coastal belt extending from Salt Range (Pakistan) in the west through south Garhwal (Uttarakhand), to Bhutan and Arunachal in the east. In the Salt Range glacigene boulder bed equivalent to that of Talchir is overlain by *Eurydesma* and *Conularia* bearing horizon which also bears *Dielasma, Fenestella* and *Aviculopecten*.

In south Garhwal, the boulder bed of Doggada is associated with slates that bear a mixed fauna that may indicate various stratigraphic levels of Permian. While *Eurydesma* and *Deltopecten* are absent, in addition to reports of forms such as *Aviculopecten* and *Fenestella* that could be of Early Permian age, the forms like *Spiriferella rajah, Lamnimargus (=Marginifera)* and *Cyclolobus oldhami* are indicative of Late Permian age. Dogadda beds are involved in imbricate thrusting and there is reason to believe that the rocks of varying ages are tectonically emplaced and derived from Tethyan facies.

In Bhutan, one of the units of Seti Khola Formation bears a diamictite horizon similar to that of Talchir and Salt Range and the carbonaceous shale associated with it bears Early Permian forms like *Trigonotreta* and *Protoretepora ampla*. In Sikkim, the Rangit pebble bed could be an equivalent of Talchir boulder bed. In Arunachal, the Lower Gondwana rocks bearing *Glossopteris* flora is underlain by fossiliferous Early Permian marine sequence referred to as Garu Formation. It bears *Dielasma, Linoproductus, Fenestella* in addition to *Eurydesma* and *Deltopecten*.

8.6.3 Kashmir

In the Tethyan part, the Permian sequence in Kashmir region is unique in that it is associated with a substantial thickness of volcanic rocks including pyroclastic material and lava. Moreover, Lower Gondwana plant beds are also locally present at different places associated with fossiliferous marine sediments similar to those known in other parts of Tethys such as Spiti, Kumaun and Nepal. The volcanic rocks are widespread throughout Kashmir valley and Pir Panjal Range, and they overlap older rocks and constitute the main topographic highs of the region. The complete sequence extending into Triassic is available from Bren spur near Nishat Bagh on the bank of Dal Lake across Zabarwan into Zewan–Khunmu–Khrew region (earlier known as the Vihi area). The sequence is also available in Liddar valley from Ganeshpura to Pahalgam though it is partially covered by recent sediments. Identical sequence is seen across Pir Panjal along Bannihal National Highway and Golabgarh pass. The general sequence is given in Table 8.1.

The Agglomeratic Slate (a misnomer as per code of nomenclature) is a polygenetic rock formation that is a result of accumulation of glacial material towards the lower part and the pyroclastic material derived from the volcanic activity re-deposited in a shallow marine and coastal environment. The association of *Eurydesma–Deltopecten* fauna is an indication of cool coastal environment as was available in Early Permian along the peri-Gondwana belt.

The volcanic activity that suddenly engulfed northern side of Indian craton appears to be a result of rifting that affected the entire Gondwana region. Due to rifting in the north, the Lhasa block was torn away from Gondwana landmass, resulting into island arc thoeliitic volcanism that continued throughout Permian and extended into Triassic. While the intensity of volcanism was maximum in Early Permian times, resulting into generation of pyroclastic material, the outpouring of lava intermittently continued through Permian and Triassic, gradually shifting its focus northwards. Accordingly, the Panjal Trap in the north Kashmir valley is of Triassic age as is indicated by the interbedded fossiliferous Triassic limestone at Imberzilwar and in Erin Valley near Bandipore. The lavas are mainly basaltic and andesitic in composition and show a thickness ranging from 400 to 1200 metres. They are greenish grey in colour, occasionally displaying glomeroporphyritic texture and are generally epidotised and chloritised. Locally the acidic varieties such as rhyolite and dacite are seen near Pantchuk and Wasterwan.

Table 8.1. Permian Sequence in Kashmir

Formation	Lithological units	Distinctive fossils
ZEWAN FORMATION	Calcareous shale with bands of calcarenite	*Cyclolobus, Xenaspis carbonaria Lamnimargus himalayensis,*
	Grey crystalline limestone	*Protoretepora ampla, Waagenoconcha, Neospirifer, Spiriferella rajah*
	-------unconformity------	
	Cream coloured slate and ash beds (*Gangamopteris* beds)	*Gangamopteris kashmirensis, Psygmophyllum, Vertebraria, Amblypterus, Actinodon*
PANJAL TRAP	Basaltic and andesitic lava flows, with rhyolitic differentiates	
	PYROCLASTIC DIVISION	
	Cream coloured slate and ash beds	Fronds of *Gangamopteris, Psygmophyllum* and *Noegerrathiopsis*
AGGLOME-RATIC SLATE	Welded tuff, volcanic ash and ignimbrite	*Eurydesma, Buxtonia, Derbyia, Trigonotreta, Warthia*
FORMATION	DIAMICTITE DIVISION Diamictite, slate and orthoquartzite	*Eurydesma cordatum, E.mytiloides, Deltopecten mitcheli, D. lyoensis, Aviculopecten*

The Zewan Formation in the type area of what was originally known as Vihi plain is mostly covered by Pleistocene Karewa rocks and alluvium with outcrops present on spurs of mountains made up of hard Panjal Trap rock. Number of these spurs notably Zewan, Risin, Ladoo, Mandakpal and Barsu comprise Zewan rocks. Unfortunately the bulk of the type formation at Zewan has been removed by quarrying for limestone during last few decades. The formation is profusely fossiliferous and various fossil cenozones and acme zones have been recognised in the formation. The formation passes upwards into Triassic without any perceptible unconformity.

8.6.4 Zanskar

In Zanskar, the equivalent of Agglomeratic Slate is a diamictite horizon that is overlain by lavas equivalent to that of Panjal Trap (referred to as Phe Volcanics). The lavas range in thickness from 230 to 300 metres, comprising 10 major flows and several minor ones. The composition is similar to that of Panjal Trap, but acidic differentiates have not been reported. The Permian rocks are involved in complicated folds together with overlying Triassic as they approach the Indus suture zone in the north. They bear Late Permian fossils—notably, productids and the characteristic index fossil *Cyclolobus.*

8.6.5 Spiti

Tethyan facies in Spiti is similar to that of Kashmir, except that there is absence of volcanic activity. In Spiti, the equivalent of Agglomeratic Slate is the diamictite horizon (referred to as Ganmachidam Formation). Unlike that of Kashmir, there are no Gondwana plant beds. This formation is directly overlain by *Eurydesma–Deltopecten* bearing formation (sometimes referred to as Gechang Formation). This formation comprises conglomerate followed by shale and ferruginous sandstone. This is followed by Kuling Shale comprising black shale, bearing phosphatic and calcareous nodules and limestone and coquina lenses. This formation bears a fauna similar to that of Kashmir—notably, *Spiriferella rajah, Waagenoconcha* and *Lamnimargus,* followed by *Cyclolobus* and *Xenaspis carbonaria.*

8.6.6 Kumaun

TheTethyan Kumaun basin is different from Spiti and Kashmir in that there is not only a total absence of Carboniferous rocks but even Lower Permian comparable to Agglomeratic Slate and its equivalents is absent. The Upper Permian rocks directly overlie Muth Quartzite with a distinct unconformity. The formation referred to as Kuling Shale comprise black shale varying in thickness from a few metres to approximately 200 meters and similar to that of Spiti. It is scantily fossiliferous but contains *Waagenoconcha, Spiriferella rajah, Lamnimargus himalayensis* and *Cyclolobus.* The shale is also known to contain some palynoflora presumed to be of Early Permian age but that is evidently reworked material from some older formation since the invertebrate fossils mentioned above are distinctly Middle and Late Permian in age. The formation passes upwards in Triassic with a perceptible break.

8.7 TRANS-HIMALAYAN PALEOZOIC SEQUENCE

Across the Indus Suture zone, the only basin within the territorial boundary of India is the Eastern Karakoram basin. Whether it was a continuation of the Tethyan basin bordering the Indian plate or a part of the Asiatic plate and thereby representing an entirely different biogeographical province is difficult to determine. Karakoram transverse fault separates the basin from the Tibetan basin to the east and Indus and Shyok suture separates it from the Ladakh granite batholith and associated Tethyan sediments of Ladakh.

The fossiliferous sequence commences from Carboniferous onwards, the older rocks being metamorphosed. The Carboniferous is represented by thinly laminated argillite interbedded with dark grey limestone and dark-grey greywacke. This formation (named as Saser Brangsa Formation) bears *Fenestella, Linoproductus, Neospirifer* and other brachiopods and bivalves. This formation has a tectonic contact with the succeeding formation (called Aqtash Formation) comprising polygenetic diamictite, chert, limestone and basic volcanics. The associated limestone bears Early Permian *Parafusulina, Pseudofusulina* and *Schwargerina.* This is the only occurrence in India where this assemblage is reported. There is also a report of plant fossils, doubtfully identified as *Glossopteris* and *Noegerrathiopsis*.

Comparing the Permo-Carboniferous sequence of Karakoram with that of Tethys Himalaya, there are some similarities. The Saser Brangsa Formation bears a fauna similar to that of the Tethyan facies. Moreover, the Aqtash Formation is reminiscent of Agglomeratic Slate and Panjal Volcanics. The plant fossils, if correctly identified, indicate Gondwana affinities. However, the presence of fusilinids is an entirely exotic element that indicates an access to the northern biogeographical province. It is interesting to note that Mesozoic fauna in Karakoram is completely different from that of Tethys Himalaya (see Chapter 10). On the basis of this evidence, there is reason to believe that Karakoram basin was a part of Lhasa block that got separated from the Indian Plate by the beginning of Permian and moved rapidly away to the north for its eventual rendezvous and collision with the Asiatic plate (see Chapter 14).

9 Gondwana Sedimentation

9.1 NOMENCLATURE AND SIGNIFICANCE

The sedimentary sequence in peninsular India is characterised by a major hiatus between the Vindhyan Supergroup and the next younger sedimentary deposition. The hiatus covers a time period extending from Cambrian to Carboniferous. The younger sequence comprises continental deposits extending in age from Early Permian to Early Cretaceous. This sequence, which constitutes one of the most characteristic and homogeneous rock groups of peninsular India, was given the name Gondwana 'System' by H. B. Medlicott way back in 1872. The name is derived from 'Gond', an aboriginal tribe of Madhya Pradesh and is a modified version of Sanskrit word *Gondwanam* meaning 'forest region inhabited by Gonds'. Since its first application, the name has become firmly entrenched in geological literature and is used in various connotations. Stratigraphically, it covers continental deposits of enormous thickness extending to several thousand metres, though the total thickness is not exposed at any one place. In course of time, Gondwana came to be associated with a particular flora which was restricted to a group of continents viz. Australia, India, Africa, Madagascar, South America and Antarctica, which were thought to be aggregated into a landmass that was initially named as Gondwanaland and subsequently came to be known as the Gondwana supercontinent (see Sec. 14.2). Apart from the similar flora, the Gondwana sedimentation was characterised by typical lithology like glaciogene sediments and universally developed coal deposits.

In the strict stratigraphic sense, Gondwana does not conform to the requirements of a formal stratigraphic grouping. Accordingly, there have been suggestions that term Gondwana be restricted as a nomenclature for the supercontinent and be used only informally for a stratigraphic grouping. It is suggested that its various divisions be accorded a formal stratigraphic status. There have been other suggestions to formalise a major part of Gondwana—except Rajmahal and Jabalpur Formations—as a Supergroup. In the succeeding pages, Gondwana is used in an informal sense as the application of this name is well understood and is an inseparable part of geological nomenclature.

9.2 DISTRIBUTION AND EXTENT

The Gondwana rocks have a widespread extension in Indian peninsula. They are distributed in isolated outcrops within the Precambrian crystalline rocks, bulk of them extending from Son valley in the north to Andhra Pradesh in the south. The thickest outcrops approximating 4,000 metres are known from Damodar valley in Bihar and Bengal, constituting the lower part of the sequence. The upper part of the sequence is best known in Damodar and Pranhita–Godavari valleys. Apart from these, the rocks occur in scores of isolated outcrops ranging in thickness from a few metres to several hundred metres. As a general rule, these rocks are found occupying basin-shaped depressions within the older formations, many of them more or less aligned to the present-day river valleys viz. Damodar, Son, Mahanadi, Pranhita, Godavari and Brahmani. They are evidently remnants of what must have been a continuous deposition, the denudation in time having removed bulk of the deposits. In most of the cases the preservation has been because of block faulting as majority of the outcrops are bounded by an abrupt line cutting them off on one side or both sides. These boundary faults are sometimes complicated and obliterated by cross faults. This faulting sometimes also extends within the basin itself, affecting the individual units of Gondwana as well.

Apart from the main distribution of Gondwana rocks in peninsular India, the equivalent rocks at various stratigraphic levels, notably towards the lower and upper part, are found associated with marine sediments. The Lower Gondwana equivalents associated with marine sediments in extra-peninsular India are known in Kashmir (see 8.6.2), Arunachal Pradesh and Assam. The Upper Gondwana rocks are associated with coastal marine sediments of Early Cretaceous age in Kutch, Tamil Nadu and Andhra Pradesh.

9.3 CLASSIFICATION

9.3.1 Historical Review

Many important units of Gondwana were already named when Medlicott proposed the term 'Gondwana'. T. Oldham had proposed the name Damuda, Mahadeva and Rajmahal, while W. T. Blanford had proposed Talchir and Panchet. Classifying them under a single grouping became problematic owing to the peculiar nature of the distribution of Gondwana rocks. As mentioned earlier, these rocks occupy basin-shaped depressions in the older formations and are usually fault bound. Accordingly, the various outcrops are isolated, and it is difficult to correlate rocks of the different isolated outcrops in light of homogeneity of lithology. There is no locality where there is a full sequence of all the units of Gondwana rocks. Some of the isolated outcrops are totally barren of fauna and flora and present difficulties in placing them in the sequence while there are many outcrops that bear plant fossils or vertebrates that help in their correlation.

9.3.2 Basis of Classification

As a general rule, Gondwana rocks are grouped into two major subdivisions based primarily on the two distinct kinds of flora. The Lower Gondwana comprises Talchir, Damuda Groups and lower part of Panchet Group, and their equivalents which are characterised by *Glossopteris-Gangamopteris* flora, whereas Upper Gondwana comprises upper part of Panchet, Mahadeva, Rajmahal and Jabalpur–Bhuj Groups and their equivalents that are characterised by *Ptilophyllum* flora. However, the boundary between the two distinct floras is not clearly marked as there is a mixing of the two floral elements in Panchet and equivalent Parsora Formation. Accordingly, the boundary has to be somewhere within Panchet and can be determined by the first appearance of *Dicroidium* flora, wherein the elements of *Glossopteris* flora may be persisting as stragglers. On account of the presence of the elements of *Glossopteris* flora these formations were generally included in Lower Gondwana for the purposes of classification, though it would not be in keeping with the principle of first appearance as a criterion in biostratigraphy. On the principle of first appearance, the rocks bearing *Dicroidium* flora should be included in Upper Gondwana even if they bear *Glossopteris* elements as stragglers. Owing to this contradiction, a three-fold classification was also proposed earlier, whereby Panchet and its equivalents were included in Middle Gondwana (Fig. 9.1).

There is a growing opinion that most of what was considered as Panchet and its equivalents should be included in Upper Gondwana and not considered as a part of Lower Gondwana as was the practice during the 20th century. These rocks can be informally considered as Panchet, as distinct from Supra Panchet rocks which are undoubtedly Upper Gondwana, while the lower part bearing *Glossopteris* flora can be considered as Panchet *sensu stricto*.

Attempts have also been made towards a biostratigraphic classification on the floral assemblages (see Table 6.3) that does away with the Lower and Upper divisions of Gondwana altogether. However, on the depositional principles the two-fold division is valid since there is an unconformity separating the two. The two-fold classification is also valid on the principle of first appearance of *Ptilophyllum* flora. Accordingly, a workable classification of Gondwana is given in Table 9.1 wherein the term Gondwana and its upper and lower divisions are informally used and only the Groups and Formations are the valid divisions.

9.3.3 Event Stratigraphic Correlation of Gondwana Sediments

As mentioned earlier, Gondwana rocks occur as graben or half-graben type (faulted on both sidesor one side only) outcrops. They usually follow the suture lines of the Precambrian basement which may have got activated during the deposition of Gondwana sediments or even after the deposition, thereby ensuring their preservation in such a setting. Historically, it has been a problem to correlate the individual outcrops in the absence of distinctive fauna and flora. In recent years, attempts have been made to correlate the widely separated and isolated patches of Gondwana units on the basis of global events marking various time slots, as observed in the entire Gondwana supercontinent. These events are based on the occurrence of glaciogene sediments, deposition of coal, association of marine fossils, transgressive and regressive phases, floral extinctions and first appearances. On these principles, seven time slots have been identified and correlated with the global time units. The classification based on these time slots is given in Table 9.2.

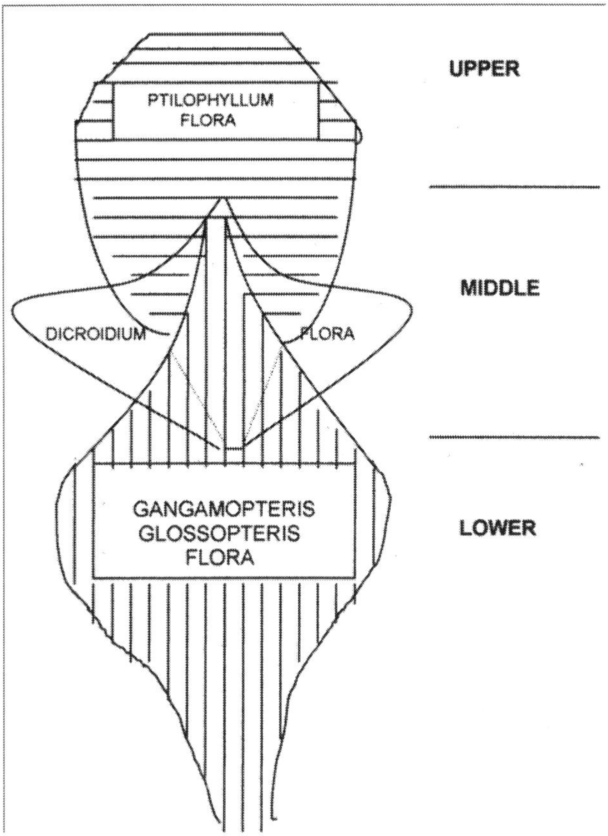

Fig. 9.1. Diagramatic representation of floral relationship of Gondwana showing mixing of flora in the middle that suggests a three-fold classification.
(Adapted and modified from Lele, 1964)

9.4 GONDWANA LIFE

9.4.1 Gondwana Flora

Gondwana flora is characterised by the dominance of pteridosperms, filicales, equisetales, cycads, conifers and cordaitales. Other elements include ginkgoales, sphenopsids and lycopsids. Pteridosperms are predominant in Lower Gondwana and are represented by *Gangamopteris, Glossopteris, Gondwanidium, Vertebraria, Samaropsis* and other related forms. These forms were flourishing throughout during the deposition of Damuda Group, covering most of the Permian Period. They started rapidly declining and most of the elements, barring *Glossopteris,* vanished by the beginning of Triassic represented by Panchet, Supra-Panchet and pre-Maleri formations. The other important elements during Permian Period include filicales *Alethopteris*, *Pecopteris*, *Sphenopteris*, equisetales *Schizoneura*, sphenopsids *Sphenophyllum*, ginkgoales *Rhipidopsis*. *Ginkgophyton*, lycopsids *Cyclodendron*, cordaitales *Barakaroxylon*, *Noeggerathiopsis* and *Dadoxylon*, the last two extending up to the end of Triassic and a little later.

Table 9.1 Classification of Gondwana Rocks

Division	Group	Type Formations	DistinctiveFauna and flora	Possible age
	Jabalpur-Bhuj	Bhuj Jabalpur	Plants: *Ptilophyllum, Pterophyllum, Nilssonia, Brachyphyllum, Pagiophyllum*	Early Cretaceous
	Rajmahal-Kota	Kota Rajmahal	*Ptilophyllum, Dictyozamites, Elatocladus, Ctenis, Nilssonia*	Jurassic
UPPER	Mahadeva	Dubrajpur Maleri Pachmarhi	*Dicroidium, Pterophyllum, Araucarites Noeggerathiopsis*	
		Supra-Panchet	Vertebrates: *C eratodus,Metaposaur, Parasuchus, Brachysuchus*	
	Panchet	Panchet	Vertebrates: *Lytosaurus, Dicynodont, Epicampodon, Gonioglyptus, Ptychognathus*	Triassic
		------------------	---Unconformity---------------------	
		Raniganj BarrenMeasures	Plants: *Glossopteris, Schizoneura, Pecopteris, Podozamites.*	
	Damuda	Barakar	*Glossopteris,Gangamopteris, Vertebraria*	
LOWER		Karharbari	*Phyllotheca, Allethopteris, Sphenopteris, Barakaria, Walkomiella, Pecopteris, Schizoneura.*	Permian
	Talchir	Talchir	*Gangamopteris, Noeggerathiopsis, Schizoneura, Gondwanidium, Phyllotheca*	

Table 9.2 Correlation of Gondwana Formations on Event Time Slots

Damodar-Koel	Rajmahal	Southern Mahanadi	Son Valley	Satpura	Godavari Valley	Time slots	Age		
	Rajmahal Trap Infra-Rajmahal		Bansa	Jabalpur	Gangpur/ Chikiala	VII	Albian Aptian	Lower	Cretaceous
Supra Panchet					Kota		Oxfordian	Upper	
	Dubrajpur	Upper Kamthi	Parsora	Upper Pachmarhi / Up. Bagra	Upper Kamthi	VI		Middle	Jurassic
								Lower	
				Lower Bagra	Dharmaram	V	Rhaetian Norian	Upper	Triassic
			Tiki	Denwa	Maleri		Carnian		
					Bhimaram	IV	Ladinian Anisian	Middle	
Panchet		Lower Kamthi	Pali	Lower Pachmarhi				Lower	
Raniganj Barren Measures		Raniganj Barren Measures	Raniganj Barren Measures	Bijori Motur	Lower Kamthi Barren Measures	III		Lopingian Guadalupian	
Barakar Karharbari	Barakar	Barakar	Barakar	Barakar	Barakar Karharbari	II	Kungerian Artinskian Sakmarian Asselian	Cisuralian	Permian
Talchir	Talchir	Talchir	Talchir	Talchir	Talchir	I			

Upper Gondwana is marked by the appearance of new pteridosperm genus *Dicroidium* that persists in Triassic and Early Jurassic. Cycads start appearing in abundance in Upper Gondwana and include *Ptilophyllum, Nilssonia, Williamsonia, Bucklandia, Otozamites* and *Dictyozamites.* Conifers are also abundant in Upper Gondwana rocks and persist in younger rocks of Juassic and Early Cretaceous age, even when cycads were on the decline. The more common conifers of Upper

Gondwana include *Elatocladus, Brachyphyllum, Pagiophyllum, Araucarites, Desmiophyllum* and *Moranocladus.*

In addition to the megaflora, a large variety of palynoflora has been recorded from Lower Gondwana rocks. On the basis of these, Lower Gondwana has been divided into several assemblage zones consisting of six varying compositions.

9.4.2 Gondwana Fauna

At the commencement of Gondwana sedimentation there were marine incursions in many parts of peninsular India (see sections 8.6.1 and 8.6.2). The marine fauna in these beds comprises elements of *Eurydesma–Deltopecten* assemblage including some productids and *Fenestella.* Fresh water fauna in Lower Gondwana comprises fishes and amphibians. The most common fish remains in Gondwana include *Amblypterus* in Lower Gondwana and *Ceratodus, Lepidotus, Dapedius* and *Tetragonolepis* in Upper Gondwana. The amphibians in Lower Gondwana include the Permian forms *Archegosaurus, Actinodon, Gondwanosaurus* and *Chelydosaurus* followed by *Gonioglyptus, Glyptognathus, Brachyops* and related forms.These get replaced in Upper Gondwana by *Pachygonia, Metaposaur* and *Mastodonsaurus* among other forms. A wide variety of reptiles are found in Panchet Formation and Maleri and Tiki Formations. Panchet forms include *Lystrosaurus, Dicynodon* and *Epicampodon* whereas Maleri and Tiki bears *Paradapedon, Parasuchus, Brachysuchus* among others.

On the basis of flora and fauna attempts have been made to biostratigraphically classify Gondwana rocks into assemblage zones. This classification is summarised in Table 9.3.

9.5 CONDITIONS OF DEPOSITION

The Gondwana sedimentation in peninsular India was initiated by deposition of glacigene Talchir sediments on an uneven Precambrian basement. The typical glacigene facies comprises tillite, conglomerate, sandstone, rhythmite and laminated shale. On a regional basis, the direction of ice movement as interpreted from till fabric appears to be from south-east to north-west. A north-westerly palaeo-drainage for the sediments is also envisaged. The Talchir sedimentation in various basins does not favour a uniform depositional model as it involves mixed facies of glacio-fluvial, glacio-lacustrine and shallow marine to tidal flat environments. At the end of glacigene sedimentation, and with the advent of a well-defined fluvial system, the Gondwana basins aerially expanded and became more or less unified into three linear basins of Koel–Damodar, Son–Mahanadi and Pranhita–Godavari–Satpura. The fluvial streams deposited essentially quartzose gravels and coarse sand constituting what is called Karharbari Formation. As the sedimentation progressed, the topography of source land located to south-east progressively became mature. The succeeding Barakar sedimentation indicates a change in the channel from braided to moderately sinuous to meandering. The resultant sediments are characterised by a progressive increase in the fine clastic facies through time from lower to upper Baraker. The overlying Barren Measures and Raniganj sediments were deposited largely by meandering streams flowing on a distinctly gentle palaeoslope directed towards north-west and west. Generally, the Triassic Gondwana sediments (Panchet and parts of Mahadeva) abound in pebbly, coarse to medium sandstone. They are suggestive of deposition by braided streams. It is also suggestive of a fresh tectonic uplift. The younger units of Gondwana have a diverse pattern in different regions and facies changes rapidly both vertically and laterally. This is so in Kota and Jabalpur Formations. Rajmahal shows association of basaltic flows interspersed and underlain with thin bands of shale rich in plant fossils and fresh water bivalves. Upper Gondwana rocks bearing flora are also found interbedded with coastal marine sediments—notably at Sriperumbadur in Tamil Nadu, Raghavapuram in Andhra Pradesh and Kutch in Gujarat. In some of these, they indicate deposition in tidal flat environment or estuarine conditions.

Table 9.3 Biostratigraphic Zonation of Gondwana Rocks

Assemblage zones	**Characteristic fossils**	**Formations**	**Possible age**
Ptilophyllum Assemblage Zone	*Ptilophyllum, Dictyozamites, Nilssonia, Elatocladus, Pagiophyllum, Brachyphyllum, Onychiopsis*	Rajmahal, Kota, Bhuj, Jabalpur, Chikiala, Gangpur, Athgarh, Bansa.	Jurassic to Early Cretaceous
Dicroidium-Neocalamites Assemblage Zone	*Dicroidium, Neocalamites, Araucarites, Schizoneura, Glossopteris, Pterophyllum, Noegerrathiopsis, Danaeopsis, Vertebraria.*	Parsora, Dubrajpur, Dharmaram	Late Triassic
Labyrinthodont Assemblage Zone	*Lystrosaurus, Epicampodon, Dicynodont, Ptychognathus, Ceratodus, Paradapedon, Brachysuchus, Metaposaur, Glossopteris, Schizoneura, Trilete* spores	Panchet, Supra-Panchet, Pachmarhi, Denwa, Almod, Mangli, Bhimaram, Maleri, Tiki	Early to Middle Triassic
Glossopteris Assemblage Zone	*Glossopteris, Barakaria, Sphenophyllum, Vertebraria, Gangamopteris, Pecopteris, Schizoneura, Cyclodendron, Allethopteris, Sphenopteris,* Microfloral assemblage covering over 50 genera of *Trilete* and *Sporite* spores	Barakar, Barren Measures, Raniganj, Bijori, Kamthi, Himgir, Motur, Pali	Middle to Late Permian
Gangamopteris Assemblage Zone	*Gangamopteris, Noegerrathiopsis, Schizoneura, Glossopteris,* Monosaccate palynomorphs notably *Parasaccites* and nonstriate disaccate forms	Talchir, Karharbari, *Gangamopteris* beds of Kashmir, Bap beds of Rajasthan.	Early Permian

9.6 LOWER GONDWANA

9.6.1 Talchir Group

The type area of this unit is in Talchir in Dhenkanal district of Orissa. This constitutes the south-east end of the Son–Mahanadi basin. The sequence here comprises approximately 300 metres of mildly dipping strata bearing glacigene boulder beds in a clay matrix at the base followed by shale-sandstone alternations. The shales are invariably olive green and splintery. In the Son valley, the Talchir rocks are associated with marine sediments representing the Early Permian transgressive phase (see sections 5.6.1 and 5.6.2). As already mentioned, they were first reported from Umaria and subsequently from Manendragarh and have since been reported from many other areas, including Damodar valley and Bap area of Rajasthan, at the same stratigraphic level. The marine sediments bear *Eurydesma-Deltopecten* fauna, characteristic of cool coastal conditions and indicative of Early Permian (Sakmarian) age. In Damodar valley, the Talchir Group commences with tillite beds followed by rhythmites, green needle shales and sandstone. The sandstone contains unaltered feldspar, indicating cool conditions. Sedimentary features are also indicative of deposition in glacial, glacio-marine and fluvio-glacial conditions. Talchir Group is approximately 500 metres thick in Satpura basin and mainly comprises diamictite, olive-green needle shale and rhythmites. Flora is usually scanty in Talchir rocks but Lower Gondwana flora consisting of *Gangamopteris* and *Noeggerathiopsis* appears for the first time in these sediments, especially in Damodar valley. The equivalents of this formation in Kashmir are *Gangamopteris* beds that underlie the marine Zewan Formation (see Sec. 8.6.3).

9.6.2 Damuda Group

The Damuda Group commences with Karharbari Formation. Earlier, Karharbari was considered to be a part of the Talchir Group, but in a number of sections, especially those in Damodar valley, there is a noticeable unconformity between the two. Moreover, there is a rapid expansion of flora in Karharbari and coal seams start appearing in abundance, distinguishing it from Talchir and makes it more akin to other members of Damuda Group. While bulk of Talchir comprises glacigene sediments, Karharbari marks the start of fluvial sedimentation. Giridih coal field is the type area for Karharbari. Maximum thickness of Karharbari is approximately 200 metres, and it comprises coarse-grained pebbly sandstone associated with carbonaceous shale and thin seams of coal. In addition to *Gangamopteris, Glossopteris, Noeggerathiopsis,* the Karharbari bears characteristic fossil genus *Gondwanidium.*

Karharbari is succeeded by Barakar Formation which is the main coal bearing formation of Gondwana. It ranges in thickness from 200 to 1100 metres and consists of cyclic sedimentation of sandstone, shale, carbonaceous shale and coal. Thick seams of coal are available in this formation in most of the coalfields of Damodar valley, Mahanadi valley and Satpura region—notably in Raniganj, Jharia, Jainti, Karanpra, Singrauli and Bokaro coalfields. Unlike Karharbari, where coal seams are thin and impersistent, Barakar coal seams are thick and laterally continuous. During

Barakar times, the basin of sedimentation considerably expanded, due to which Barakar sediments overlap the basement rocks in most of the basins. The Barakar sandstone is commonly channel shaped in lower part and tabular sheet-like in upper part. Leaf and stem impressions are highly abundant in Barakar rocks especially in carbonaceous shale and coal. In addition to the abundance of *Glossopteris*, the genera *Sphenophyllum, Barakaria* and *Walkomiella* appear in this Formation.

Barren Measures that succeed Barakar Formation are so named because they are devoid of coal. The lithology comprises alternating units of cross-bedded ferruginous sandstone, micaceous siltstone and ferruginous shale. Owing to this composition, the formation is referred to as Ironstone Shale in Raniganj coalfield and Wardha-Godavari valley. The contact with underlying Barakar and overlying Raniganj is usually gradational. The flora is poor in this formation and comprises fronds of *Glossopteris* and *Cyclodendron.*In the Satpura region, the equivalent of this unit is named Motur Formation. The lithology of Motur Formation comprises red clay, feldspathic sandstone and grey shale. It also bears a few impersistent lenticles of coal.

Raniganj Formation overlies Barren measures and has a widespread distribution throughout the Gondwana terrain, though it is variously named in different basins. It is thickest in the Koel–Damodar basin where it attains a maximum thickness of over 1000 metres. It comprises coarse to medium sandstone fining upwards. The sandstone is interbedded with shale and seams of coal. The sandstone is channel to tabular shaped with abundant planar and trough cross beds. Coal seams are thin to moderately thick and are laterally persistent for several kilometers. The only thick coal seam is the one present in Singrauli coal field (Jhingurdah coal seam) which is approximately 160 metres. Raniganj marks the acme of *Glossopteris* flora. Two species of *Glossopteris*—namely, *G. conspicua* and *G. retifera* are the most abundant and characteristic forms of this formation—in addition to a wide variety of filicales, equisetales and pteridosperms. In the Wardha–Godavari basin, the equivalent of Raniganj Formation is known as *Kamthi Formation*. The thickness of this formation ranges from 40 to 400 metres and it comprises conglomerate, conglomeratic sandstone, siltstone and shale. In Satpura region the equivalent unit is the Bijori Formation that is estimated to be approximately 1600 metres in thickness. This formation is characterised by the occurrence of the famous labyrinthodont species *Gondwanosaurus bijoriensis.* In the Mahanadi valley, a diverse assemblage of rocks overlapping the Barakar Formation is considered to be the equivalent of Raniganj and is referred to as Himgir Formation.

9.7 UPPER GONDWANA

9.7.1 Panchet Group

Panchet Group is best exposed in Panchet Hill in the southern part of Raniganj basin in Damodar valley. It is not exposed in Jharia basin but is known in several other basins of Damodar valley. The rocks unconformably overlie the Damuda Group. The formation comprises greenish, buff and brown, current bedded sandstone, shale and siltstone. Maximum thickness is approximately 600 metres. The flora is scanty and comprises *Schizoneura, Glossopteris and Cyclopteris*, in addition to a wide variety of vertebrates notably *Lystrosaurus, Epicampodon, Glyptognathus, Ptychognathus* and *Indobrachyops.* The doubtful equivalent rocks of Panchet in Satpura and the Wardha–Godavari valley are referred to as Almod and Mangli beds, respectively.

9.7.2 Mahadeva Group

The name Mahadeva was given by Oldham in 1856 to a sequence of unfossiliferous rocks exposed in Mahadeo hills area near Pachmarhi. Since then, the name has been widely used to designate the generally thick and poorly fossiliferous unit of Upper Gondwana. Supra–Panchet rocks which were considered earlier as a part of Panchet Group are now considered as equivalents of the basal units of Mahadeva Group. They have a maximum development in various basins of Damodar valley—notably in Bokaro basin where they attain a thickness of approximately 600 metres. The SupraPanchet rocks rest on Panchet rocks with a distinct unconformity. The lithology consists of ferruginous sandstone, conglomerate with bands of ferruginous siltstone. SupraPanchet rocks are, by and large, unfossiliferous, and, accordingly, it is difficult to correlate them with the rocks of other basins. Supra–Panchet happens to be the youngest Gondwana unit present in Damodar valley. The bulk of Upper Gondwana sediments are exposed in Rajmahal Hills, Satpura region and Pranhita–Godavari valley. For the Mahadeva Group, the type area is the Satpura region where it has a maximum development.

In Satpura basin Mahadeva Group consists of Pachmarhi, Denwa and Bagra Formations. Pachmarhi comprises thick beds of white sandstone inter-layered with pebble beds and has a maximum thickness of approximately 750 metres. No fossils have been reported from this formation. It is conformably overlain by Denwa Formation which comprises sandstone alternating with clay bands and has a thickness of approximately 300 metres. Denwa Formation is known to contain plant fossil *Phoenocopsis* and labyrinthodonts *Mastodonosaurus indicus, Metaposaur* and *Paratosaurus.* Denwa Formation has a vertical and lateral transition into Bagra Formation comprising conglomerates. The thickness of the latter varies from 200 to 240 metres.

In the Pranhita–Godavari valley, the equivalents of Mahadeva constitute Bhimaram, Maleri and Dharmaram Formations. Bhimaram is an unfossiliferous unit comprising sandstone with intercalated red clays and is comparable to the Pachmarhi Formation. It is overlain by Maleri Formation that constitutes approximately 250-metre-thick sandstone, red clay and lime-pellet rocks. This formation is rich in vertebrate remains—notably, fishes, labyrinthodonts and reptiles. The more abundant elements include species of *Ceratodus, Metaposaur maleriensis, Paradapedon huxleyi* and *Brachysuchus maleriensis.* Dharmaram Formation that overlies Maleri Formation comprises gritty sandstone interbedded with red clay. The formation bears fragments of prosauropod dinosaur bones that compare well with *Plateosaurus.*

In the upper reaches of the Narmada valley in the southern Rewa region, the equivalent of Maleri and Denwa Formations is referred to as Tiki Formation owing to its exposure near Tiki. It also bears a rich vertebrate fauna similar to that of Maleri. Of special interest in this area is the Parsora Formation which is associated with Tiki Formation. This Formation comprises ferruginous sandstone, mottled shale and ironstone. This formation has generated a lot of controversy owing to the presence of mixed flora in it namely the association of *Glossopteris* elements of Lower Gondwana affinities and *Dicroidium* of Upper Gondwana. Owing to this, admixture many authors previously recommended a three-fold classification of Gondwana.

The representative of Mahadeva group in Rajmahal Hills is known as Dubrajpur Formation. It shows a wide areal extent in the western part of Burdwan district in West Bengal and consists of feldspathic sandstone, red, green and carbonaceous shale with occasional lenses of coal. It also bears *Ptiliphyllum* and *Gleichnites* towards the upper part of the formation

9.7.3 Rajmahal–Kota Group

The Rajmahal Formation is best exposed in Rajmahal Hills which is situated in north-eastern part of Jharkhand. The group overlies Dubrajpur Formation that has a widespread extent. This is a characteristic Group of Upper Gondwana since it contains typical upper Gondwana flora in abundance. The bulk of the Group in Rajmahal hills comprises Rajmahal trap that is interbedded with Rajmahal intertrappean shale beds. The trap comprises olivine basalt flows which have a limited extent that may have been widely distributed but have since been eroded. The traps with their interbedded sediments are divided into four levels viz. A, B, C, D, on the basis of lithology and the presence of flora in the intertrappean beds in each level. The intertrappean beds in levels A and D are unfossiliferous or scantily fossiliferous. However, intertrappean beds in levels B and C are profusely fossiliferous. The fossils comprise typical elements of Ptillophyllum flora notably *Ptillophyllum, Pterophyllum, Nilssonia, Brachyphyllum, Pagiophyllum* and *Elatocladus*.

In the Pranhita–Godavari valley, Kota Formation may be the equivalent of Rajmahal Formation. This formation derives its name from the village Kota which is situated on the east bank of river Pranhita to the north of its confluence with Godavari. The formation unconformably overlies the Dharmaram Formation. The formation commences with a pebbly sandstone bed that passes gradually upwards into fine white sandstone, red clay and beds of limestone, ferruginous mudstone and siltstone. It contains fishes (*Lepidotes, Dapedium*) and sauropod remains in addition to typical Upper Gondwana floral elements like *Elatocladus, Otozamites* and *Pagiophyllum.*

In Mahanadi valley in Orissa, the Upper Gondwana is represented by 400 metres thick formation comprising conglomerate, grit and variegated shales known as Athgarh Formation. This formation is equivalent of Rajmahal Formation and contains typical Upper Gondwana flora, notably *Ptilophyllum, Dicroidium* and *Otozamites.*

9.7.4 Jabalpur-Bhuj Group

Overlying the Bagra Formation in Satpura region there is about 250 metres thick formation comprising conglomerate, clay and cross bedded sandstone that is rich in

Upper Gondwana flora. This formation is known as the Jabalpur Formation owing to its exposure around the town of Jabalpur. It contains only a few remains of cycads but conifers such as *Pagiophyllum* and *Brachyphyllum* are more abundant. Besides, the form *Onychiopsis* appears for the first time in this Formation.

Comparable to Jabalpur Formation and interbedded with coastal marine Early Cretaceous rocks in Kutch is another Upper Gondwana unit known as Bhuj Formation (see Sec. 10.5.2). Bhuj Formation consists of sandstone that displays evidence of lagoonal or estuarine deposition in a tidal flat environment and appears to be bioturbated. The flora is similar to that of the Jabalpur Formation with abundant elements of conifers such as *Brachyphyllum* in addition to *Ptilophyllum, Williamsonia, Elatocladus* and *Onychiopsis.*

■■■■■

10 Mesozoic Marine Sequence

10.1 PERMIAN-TRIASSIC BOUNDARY

10.1.1 Global Concept of Permian – Triassic Boundary

The Permian–Triassic boundary (P/T boundary) is associated with the largest extinction event in Phanerozoic time. An entire lot of marine organisms which abounded the Paleozoic seas became extinct within a span of 5–10 million years. It is estimated that nearly 93–95% marine species that existed in Late Permian were wiped out. However, the percentage loss of land and fresh water organisms was comparatively less; thereby, the overall percentage extinction was of the order of 70%–75%. This brought about a total transformation in the biological scenario and Mesozoic life was almost entirely different from that of Paleozoic. The termination of Permian marked an evident environmental catastrophe that was not only reflected in faunal and floral extinctions but also in a pronounced negative carbon and strontium isotope anomalies and a positive sulphur isotope excursion. There was a large fall in $\delta^{13}C$ content of the surface ocean probably due to a rapid drop in the sea level. The salt content of the ocean also dropped by approximately 4% and pronounced anoxic conditions existed in most of the oceans. The end of Permian also marked an increase in aridity in the continental areas. There was also a phase of sub-aerial volcanism in Siberia and China.

The boundary between Permian and Triassic cannot be exactly fixed in most of the marine sequences across the globe due to several reasons. For one, as mentioned above, the closing stages of the Permian were characterised by one of the strongest regressions of the sea in the entire Phanerozoic. Due to this, the boundary beds are missing in most of the marine sections of the world. Secondly, the extinction has been of a stepped type extended over a prolonged time period. Consequently, in several sections, we find a zone of mixed fauna where Paleozoic elements extend and are found together with younger Triassic elements. While it is easy to recognise the final phase of Permian, it is difficult to identify the exact beginning of Triassic, since it happens to be at different levels in the known marine Permo-Triassic sequences. This is because of a break in sequence due to the regression of the sea and consequent break in deposition.

The Stratigraphic Commission has recognised the sequence at Meishan, Zhejiang Province, southern China as the GSSP for the base of Triassic (see Appendix I). It coincides with the first appearance of the conodont *Hindeodus parvus* (equivalent to what was originally described as *Isarcicella parva*). This is the first zone for the cosmopolitan correlation of the beginning of Triassic following the end of Permian mass extinction. This conodont species appears just after the lowest level spike of carbon isotope indicative of a minimum biological abundance. All the environmental imbalances including isotope anomalies are thereby assigned to the latest Permian. On land, this conodont event is close to the disappearance of typical Permian *Dicynodon* tetrapods after an interval of co-occurrence with Triassic dicynodont *Lystrosaurus*. The boundary can be dated to 251.0 ± 0.4 million years.

The cause of faunal and floral extinction at the Permian-Triassic boundary level is still a matter of speculation. In another major extinction at the Cretaceous–Tertiary boundary level, the iridium anomaly points towards the impact of an extra-terrestrial object (see Section 11.6). However, there is no perceptible iridium concentration at the P/T boundary level. Accordingly, a bolide impact does not appear to be a cause of the environmental catastrophe that gripped the globe at the end of Permian. It is evident that several factors may have contributed towards the ecological imbalance that brought the organisms to extinction in a stepped manner. A common hypothesis to explain the cause of ecological imbalance and disturbed climate is that the release of aerosols and greenhouse gases generated by Siberian and Chinese continental flood basalts brought about greenhouse warming alternating with cold volcanic winter episodes.

10.1.2 Permian–Triassic Boundary in India

In the marine sequences of the Indian sub-continent, the Permian passes upwards into Triassic with an apparent conformity in a number of basins which notably include Spiti–Zanskar, Kashmir and Salt Range of Pakistan. The horizon bearing the distinctive conodont *H. parvus* has not been reported in any of these sections except for a doubtful mention in Spiti. Instead, the earliest fossil horizon of definite Triassic age is the zone bearing the ammonoid *Otoceras woodwardi.* This ammonoid zone is slightly higher than *H. parvus* zone which has been identified as the earliest zone of Triassic. The absence of the latter in Himalayan sequences can either be facies related or a result of a minor diastem because, otherwise, there does not appear to be any significant break at the boundary level.

One of the best sections in India where the Permian–Triassic boundary can be easily identified and which is also easily accessible is the Guryul Ravine section in Kashmir. This section is approximately eight kilometres from Srinagar in what was originally named as the Vihi area and is extremely close to the type section of Permian Zewan Formation. This section was also submitted to the International Stratigraphic Commission for consideration as an additional GSSP, but it has not been accepted as such. The Guryul succession consists of arenaceous sequence bearing typical Permian fauna notably *Cyclolobus* and *Lamnimargus* (=*Spinomarginifera*), followed by black shale that grades upwards into carbonate rocks. The black shale bears Permian brachiopods (*Lamnimargus*) in association with typical Early Triassic bivalve *Claraia*. The zone of mixed faunas is a characteristic feature at the boundary level and is also known in Pahalgam and Pastun. It is also reported from Salt Range in Pakistan. The Permian elements are evidently the stragglers which may have survived for a short while before final extinction. However, the boundary must be at the base of the black shale as it is the first appearance and not the final extinction that determines the start of a new Period. For Indian marine sequences, *Otoceras woodwardi* in association with *Claraia griesbachi* is considered as marker for the beginning of Triassic and this level is only marginally higher than the level of *H. parvus* that is globally considered as the boundary marker.

On land, the boundary between Permian and Triassic in the Gondwana sequence appears to be at the base of Panchet Formation or somewhere within it as this formation and its equivalents bear *Lystrosaurus*. In the Raniganj basin the Panchet Formation commences with khaki green micaceous siltstone bearing fossil estherids, a group of crustaceans that are characteristic of Early Triassic. The underlying beds bear *Glossopteris–Schizoneura* flora, while the overlying beds bear *Lystrosaurus*. The estherid horizon is considered to be the beginning of Triassic and may be recognised as continental stratotype of P/T boundary.

10.2 MESOZOIC SEDIMENTATION AND CLIMATE

The beginning of Triassic has been a period of global regression in sea level due to which large tracts of land were exposed to sub-aerial denundation. This was also a period of arid climate in the Gondwana supercontinent. Most of the Gondwana region was exposed to sub-aerial weathering, and the sea had receded. Conseqently, marine sedimentation was restricted to the Tethys basin alone with no marine incursions into Lesser Himalaya or peninsular India, barring Salt Range region. The warming up in Late Triassic and following Jurassic and Cretaceous Periods caused the inundation of coastal areas of Indian peninsular landmass, due to which marine sedimentation took place on peninsular coastline in Kutch, Rajasthan and eastern coast of south India. There were several transgressive and regressive cycles in this coastal sedimentation that were more or less co-eval with the global eustatic. However, in extra-peninsular region to the north, the transgressive and regressive phases were determined by the northward movement of the Indian plate and corresponding oscillations in the continental shelf, and were not always co-eval with the global transgressive and regressive cycles. Thus, it is difficult to correlate the marine Jurassic and Cretaceous sequence of Himalaya with that of peninsular India since the phases of transgression are not identical.

On a global level, Triassic was a period when the surface temperature was increasing at a rapid rate and carbon dioxide content in the atmosphere had almost trebled from the beginning of Triassic to that of Jurassic. Thus, the climatic conditions on the Earth were going through radical changes. It is estimated that the carbon dioxide content in the atmosphere at the beginning of Jurassic was at the highest level and almost 10 times that of the present day. There was no ice on the poles by the end of Triassic and beginning of Jurassic. During the Jurassic period, the aridity was replaced by a humid climate but the temperature was still high and almost greenhouse conditions prevailed. The high temperature resulted in gradual rise in sea level, but heavy precipitation generated lush forests with profuse vegetation and coal formation. Towards the end of Jurassic, there was gradual cooling and by Early Cretaceous cool temperate forests covered the polar regions. However, the temperature was still high compared to present-day and no ice existed on the poles throughout Cretaceous period. There are, however, evidences of intermittent phases of glaciation in different parts of the world that did not last for long periods.

10.3 TRIASSIC AND EARLY JURASSIC SEDIMENTATION IN HIMALAYA

10.3.1 Distribution

As has been mentioned earlier, (see Sections 6.3 and 10.2) Triassic and Early Jurassic sedimentation was restricted to Tethys Himalaya and Salt Range of Pakistan only. The global regressive phase had only a marginal effect in Tethys region because of the northward movement of the Indian Plate and the resultant subsidence of the cratonic margin. There was a continuous sedimentation throughout Triassic in Tethys basins of Kashmir, Spiti–Zanskar and Kumaun-Nepal. Towards the later part of Triassic and beginning of Jurassic, there was a shallowing and rejuvenation of the cratonic margin that eventually brought the sedimentary cycle to an end. Consequently, the Triassic sedimentation along Tethys is one of the best and a completely preserved sequence in Higher Himalaya.

There is another facies of Triassic sedimentation to the north of Indian craton along the Indus Suture Zone. This constitutes the flysch facies and is popularly referred to as the Indus flysch. It comprises what has been named as the Lamayuru Formation that bears flyschoidal rocks in which are embedded several olistoliths (exotic blocks) of varying ages, notably Permian.

10.3.2 Kashmir

Triassic sediments are extremely well exposed in mountains surrounding Kashmir valley along the eastern and southern sides. One of the best and nearly complete sequences of these rocks is present in the Vihi area from Khunamuh, which includes the Guryul Ravine section (see Section 10.1.2 above), extending across into the mountains bordering the northern and eastern side of Tral valley. Triassic rocks extending into Early Jurassic are also well exposed along the northern slope of Pir Panjal Range from Liddar valley to Verinag and in the east from Kolhai through Amar Nath Cave, Baltal to across Zojila mountain range.

Lithologically, bulk of the sequence comprises carbonate rocks with which are interbedded shales, especially in the middle part. The main outcrops along Vihi area and Tral valley can be lithostratigraphically divided into four formations: Khunamuh Formation, Tindra Formation, Khreuh Formation and Wuyan Formation. Of these, Khunamuh and Tindra Formations are of Early Triassic age whereas Khreuh and Wuyan Formations are of Middle and Late Triassic ages, respectively. Early Jurassic (Hettangian) rocks are exposed along the northern slope of Pir Panjal Range near

Verinag and constitute Wumuh Formation, based on their exposure near village Wumuh above Verinag and below the northern mouth of Jawahar Tunnel. Equivalents of this formation are also known in Warwan valley along the south-eastern slope of Pir Panjal Range (Drumgam Formation) and in Baltal and across Zojila mountain at Matayin.

Table 10.1 Triassic Sequence in Tral Valley (Kashmir)

FORMATION	LITHOLOGY	FAUNA	THICKNESS	AGE
Wuyan Formation	Limestone	Crinoid stems *Montlivaltia* sp., *Stylophyllum* sp. *Myophoria* sp. *Spiriferina* sp.	280 m	Late Triassic
Khreuh Formation	Calcarenite Shale alternating with calcarenite	*Daonella indica*, *Daonella lamelli* *Avicula* sp., *Lima* sp. *Ceratites* sp. *Ptychites* sp. *Gymnites* sp., *Orthoceras* sp.	240 m	Middle Triassic
Tindra Formation	Nodular limestone	*Keyserlingites* sp., *Neospathodus timorensis*, *Prohungarites* sp. *Hungarites* sp., *Flemingites* sp., *Neospathodus triangularis*	80 m	Early Triassic
Khunamuh Formation	Nodular Limestone Calcarenite Limestone	*Neospathodus homeri*, *N. spathi*, *N waageni*, *Gondolella nepalensis* *Meekoceras* sp. *Neospathodus kummeli*, *N. dieneri* *Ophiceras sakuntala*	140 m	Early Triassic

As discussed above (see section 10.1.2) the earliest fossil horizon in Kashmir Triassic bears *Claraia griesbachi* followed by *Ophiceras sakuntala* zone. The beds bearing *Claraia griesbachi* also contain Late Permian forms like *Lamnimargus (=Spinomarginifera)* and some productids. This is the horizon bearing mixed fauna wherein some elements of Permian have straggled into the beginning of Triassic. The boundary between Lower and Middle Triassic is just above the horizon bearing the ammonite *Keyserlingites* and conodont *Neospathodus timorensis.* As a general rule, fossils are profuse in Lower Triassic and gradually decline in numbers upwards. The Upper Triassic namely Wuyan Formation comprises hard grey massive limestone which is scantily fossiliferous. Triassic sequence in type area of Tral Valley is given in Table 10.1.

10.3.3 Spiti–Zanskar

Overlying the Permian Kuling Shale, an almost complete and fossiliferous Triassic sequence extending into the beginning of Jurassic, is exposed in Spiti Valley that is referred to as Lilang Group. This group was studied in detail by Hayden, and the fauna determined by Diener towards the beginning of 20th century, and the classification proposed by them has stood the test of time. Since this classification is not in keeping with the code of stratigraphic nomenclature that was proposed much later, in recent years, attempts have been made to superimpose a lithostratigraphic division on this classification. Table 10.2 gives a summarised version of both these classifications.

Table 10.2 Classification of Lilang Group in Spiti

Informal Nomenclature of beds	Lithology	Characteristic Fossils	Thickness	Epoch	Age	Form-ation
Tagling Limestone	Massive Limestone and Dolostone	*Megalodontoids, Thecosmilia*	500 m	Early Jurassic	Hettan-Gian	Kioto
Megalodon Limestone	Massive Limestone and Dolostone	*Megalodon ladakhensis*	250 m	Late Triassic	Rhaetian	
Quartzite 'series'	Ortho-quartzite, Grey limestone, Black shale	*Spirigera maniensis*	100 m			Nunuluka
Monotis Shale	Shale,Shaly limestone, Calcarenite	*Monotis salinaria, Spiriferina griesbachi*	100 m		Norian	Alalor
Coral limestone	Limestone	*Spiriferina griesbachi, Rhynchonella bambanagensis*	30 m			Hangrang
Juvavites beds	Brown shale, Limestone, Sandstone	*Juvavites angulatus*	150 m			Sanglun
Tropites beds	Dolostone, Limestone, Shale	*Tropites subbullatus, Dielasma julicum*	300 m		Carnian	
Grey beds	Grey shale, Shaly limestone	*Spiriferina shalshalensis, Joannites cymbiformis*	150 m			

<table>
<tr><td>Halobia beds</td><td>Dark splintery limestone</td><td>Halobia cf. comata, Joannites thananensis</td><td>45m</td><td></td><td></td><td>Chomule</td></tr>
<tr><td>Daonella limestone</td><td>Hard dark limestone</td><td>Daonella indica</td><td>45m</td><td rowspan="6">Middle Triassic</td><td rowspan="2">Ladinian</td><td rowspan="2">Kaga</td></tr>
<tr><td>Daonella shale</td><td>Shale, Shaly limestone, Black limestone</td><td>Daonella lammeli, Ptychites gerardi</td><td>50m</td></tr>
<tr><td>Upper Muschelkalk</td><td>Concretionary limestone</td><td>Ptychites rugifer</td><td>6m</td><td rowspan="4">Anisian</td><td rowspan="7">Mikin</td></tr>
<tr><td>Lower Muschelkalk</td><td>Shale, Dark shale, Grey limestone</td><td>Keyserlingites dieneri, Sibirites prahalada, Spiriferina stracheyi</td><td>2m</td></tr>
<tr><td>Nodular limestone</td><td>Dark nodular limestone</td><td>Scanty indeterminate fossils</td><td>20m</td></tr>
<tr><td>Basal Muschelkalk</td><td>Shaly limestone</td><td>Rhynchonella griesbachi</td><td>1m</td></tr>
<tr><td>Hedenstroemia beds</td><td>Limestone, Shale</td><td>Hedenstroemia mojsisovicsi, Flamingites rohilla, Xenodiscus nivalis</td><td>182m.</td><td rowspan="3">Early Triassic</td><td>Olenekian</td></tr>
<tr><td>Meekoceras zone</td><td>Thin bedded limestone, Shale</td><td>Meekoceras varaha, M. lilangense</td><td>1 m.</td><td rowspan="2">Induan</td></tr>
<tr><td>Ophiceras zone

Otoceras zone</td><td>Grey limestone

Thin bedded limestone</td><td>Ophiceras sakuntala, Claraia griesbachi

Otoceras woodwardi</td><td>3 m.</td></tr>
</table>

As is evident from the table, the Lilang Group comprises a dominant component of carbonate rocks of enormous thickness. The rocks are widely distributed through Spiti valley, Lahaul and Zanskar area extending as far as the boundary of the Indus Suture in Ladakh. It overlies the Permian Kuling Shale with an apparent conformity and is overlain by Upper Jurassic Spiti Shale.

10.3.4 Kumaun–Nepal

In the Kumaun–Nepal region the Triassic rocks overlie the Kuling Shale with a sharp contact. The rocks are widely distributed extending from Niti pass section through Rimkhim along Shalshal cliff, Girthi Ganga and Kali river valleys. The thickness is much less compared to that of Spiti–Zanskar and does not exceed 800 metres. Lithologically, the sequence is divisible into four formational units namely Chocolate Formation, Kalapani Limestone, Kuti Shale and Kioto Limestone. While the upper part of the sequence namely Kioto Limestone, though somewhat truncated, closely corresponds to that of Spiti and is scantily fossiliferous, the lower three units are rich in fossils. The Chocolate Formation bears *Otoceras, Ophiceras* and *Meekoceras* zones and is of Early Triassic age. Kalapani Limestone comprises massive nodular limestone with a few shale alternations towards the upper part. It is rich in fossils which notably include *Keyserlingites dieneri, Sibirites prahalada, Spiriferina stracheyi, Ptychites rugifer, Rhynchonella griesbachi* and is of Anisian age. Kuti Shale bears *Daonella indica, Halobia* cf. *comata, Spiriferina shalshalensis* and is of Ladinian and Carnian age. Norian part of the sequence is truncated and represented by a few metres of limestone shale alternations at the base of Kioto Limestone that bear *Spiriferina griesbachi*. Kioto Limestone bears *Megalodon* horizon but, is by and large, unfossiliferous.

10.3.5 Indus Valley (Ladakh)

An enormous thickness of flyschoidal rocks exposed along the Indus suture zone between Mulbekh through Lamayuru to Khaltse constitutes Lamayuru Formation that is also popularly referred to as Indus Flysch. Lamayuru Formation comprises turbiditic sandstone, siltstone, carbonaceous shale and thin bands of limestone. The Formation is studded with olitholiths (exotic blocks) of various dimensions ranging up to a maximum of 80 metres, consisting mostly of limestone. The formation is essentially of the nature of wildeflysch and commences with a thick zone of turbidites and graded bedded calcarenite.

The age of Lamayuru Formation has been a debatable issue. It has been assigned ages ranging from Triassic to Cretaceous. The problem has been the paucity of faunal data in the main formation, while most of the exotic blocks are fossiliferous. The blocks range in age from Late Permian to Carnian and are evidently derived from the Tethys basin to the south. Most of the faunal data indicating a Cretaceous or even Paleocene age is doubtful and probably derived from younger rocks. There is reason to believe that bulk of the Lamayuru Formation is the facies variant of Triassic–Liassic sequence of Tethys, as no exotic blocks younger in age than Carnian are found in the rocks in the Lamayuru Formation *sensu stricto*. However, associated with Lamayuru Formation are mélange rocks that may be of Cretaceous age and are usually grouped with Lamayuru Formation since they are also flyscoidal in nature. Lamayuru Formation was evidently deposited due to the subsidence of the northern margin of the Indian craton following the separation of Lhasa block from the Indian plate (also see section 6.3.1).

10.4 JURASSIC-CRETACEOUS SEQUENCE IN HIMALAYA

10.4.1 Distribution

Jurassic–Cretaceous transgressive-regressive cycle that starts in post-Hettangian times comprises a series of sub-cycles in Himalaya. Following the Hettangian stage, there is a general break in sequence throughout Himalaya. In the Kashmir basin, there is no marine sedimentation after that stage. In the Spiti–Zanskar basin the break extends upto the beginning of Oxfordian, covering later part of Early Jurassic and whole of Middle Jurassic. In Kumaun and Nepal, the break extends at least up to Bajocian, following which there is punctuated sequence covering parts of Bathonian and Callovian succeeded by a continuous succession from Oxfordian extending into Cretaceous. The transgressive phase commencing from Oxfordian extends right through Late Jurassic and the sedimentation is of a deep outer shelf facies. In Early Cretaceous, there is a shallowing of the basin. The two facies have a vertical and lateral variation. Towards the later part of Cretaceous there is a rapid deepening of the basin. On the northern margin of the Tethyan basin—notably in Kumaun in Malla Johar region—deep marine oozes are associated with ophiolitic mélange bearing huge exotic blocks of older age (Triassic) derived from northern Tibetan region.

Along the Indus Suture Zone the Mesozoic rocks are represented, apart from Lamayuru flysch, by a thick suite of volcanic rocks associated with sediments, ophiolites and ophiolitic mélange. These rocks are structurally associated with sediments of Early Paleogene age along a series of thrusts constituting the Suture Zone.

10.4.2 Spiti

Spiti Formation (also referred to as Spiti Shale) overlies Kioto Limestone of Rhaetian–Hettangian age with a sharp disconformable contact. This formation has a characteristic feature because of its dark colour. It comprises dark grey to black splintery and friable shale, intercalated with extremely few bands of calcareous sandstone, especially towards the top. The shale bears a profusion of phosphatic, cherty and calcareous nodules that are often nucleated by fossils of various sizes. The formation is approximately 300 metres thick and has a widespread extension throughout Spiti basin and sporadic distribution in south-eastern part of Zanskar. Spiti Formation is profusely fossiliferous bearing a wide range of cephalopods and bivalves. The most common forms are the belemnite species *Belemnopsis gerardi* and the ichnogenus *Zoophycos*. Seven assemblage zones namely, *Mayaites, Torquatisphinctes, Pachysphinctes, Aulacosphinctoides, Spiticeras, Virgatosphictes* and *Himalayites* have been recognised in the sequence. On the basis of fauna, the age ranges from Oxfordian to Tithonian, covering the total range of Late Jurassic. The deposition must have taken place in mid-shelf region through a slow rate of sedimentation with occasional mudflows and tubidity currents.

Giumal Formation (also called Giumal Sandstone) conformably overlies Spiti Formation when the latter passes imperceptibly upwards into it with the increase in sandstone bands. Giumal Formation comprises essentially sandstone and glauconite bearing sandstone bands, siltstone and occasional gritty sandstone bearing pebbles. The sandstone ranges from quartzarenite to quartzwacke and is poorly sorted. The thickness is over 200 metres. Compared to Spiti Formation, Giumal Formation is sparsely fossiliferous. Faunistically, the boundary between Spiti Formation and Giumal Formation closely corresponds with the Jurassic- Cretaceous boundary. The Tithonian ammonites—namely, *Virgatosphictes, Himalayites, Spiticeras* and *Aulacosphinctes*—suddenly vanish and are replaced by Berriasian forms

Neocosmoceras, Berriasella and *Odontodiscoceras.* Deposition of the formation in the Spiti area appears to be in a shallow shelf.

Chikkim Formation overlies Giumal Formation with an apparent conformity. The Lower Chikkim Member comprises limestone whereas the Upper Chikkim Member comprises shale. The total thickness of Lower Member is approximately 55 metres, while that of upper Member is more than 30 metres. From the basal part of Lower Member planktonic foraminifers *Bitcinella breggiensis* and *Planomalina buxtorfi* have been recovered that indicate an Albian age. *Whiteinella archaeocretacaea* at a level approximately 26 metres above the base indicates Cenomanian/ Turonian boundary. The Upper Member, in addition to belemnites, irregular echinoids and benthic rotalids, bears various species of *Globotruncanita* and *Gansserina* indicating a Campanian age. The total range of Chikkim Formation could, therefore, extend from Albian to Campanian and may even have extended into Maastrichtian. While the Lower Member could be deposited in mid-shelf conditions, the Upper Members would have been deposited in outer shelf or along the shelf margin.

10.4.3 Zanskar

Jurassic–Cretaceous rocks of Zanskar are somewhat different from those of Spiti. For one, Spiti Formation has a limited thickness and distribution here. It is sporadically distributed and has a thickness of only 20–30 metres. Giumal Formation ranges in thickness from 250 t0 350 metres and comprises lower, thick bedded coarse sandstone unit followed by dark grey silty shale intercalated with grey to green sandstone. Chikkim Formation is approximately 90 metres in thickness and comprises dark limestone towards the lower and upper part with a thick component of dark shale in the middle. The other difference from that of Spiti is that Chikkim Formation is overlain by a sequence of flyschoidal rocks covering a part of Upper Cretaceous (Kangi La Formation). It contains planktonic forminifers—namely, *Heterohelix globulosa*, *Pseudotextularia elegans* and *Globotruncana*—together with turrilitid ammonites that indicate a Campanian age. Kangi La Formation itself is overlain towards the eastern side by rocks of Maastrichtian age (variously named as Spanboth Formation or Marpo Limestone) containing orbitoid foraminifer *Omphalocyclus macroporus* and *Siderolites calcitrapoides*. Both these rock units are evidently the facies variants of the upper part of Chikkim Formation of Spiti as they indicate the same age. These rocks further pass upwards into Paleocene sequence.

10.4.4 Kumaun and West Nepal

In Kumaun, Kioto Limestone is overlain unconformably by 60-metre-thick Lapthal Formation comprising thin bedded limestone with bands of coquina. The shells constituting the coquina are generally indeterminate, but some identifiable ammonites such as *Macrocephalites* and *Dolikocephalites* and palynomorph taxa suggest a possible Late Bathonian to Callovian age. There is another break following Lapthal Formation when a thin band of ferruginous oolite approximately three metres thick unconformably overlies this Formation. This unit is continuosly deposited throughout Kumaun and western Nepal and is sometimes also referred to as Salcacutus bed because of the presence of *Belemnites salcacutus.* Spiti Formation similar to that of Spiti unconformably overlies this bed and is approximately 300 metres thick. The overlying Giumal Formation comprising grayish green sandstone and glauconitic sandy shale bears bands of radiolarian chert indicative of the rapid deepening of the basin. The equivalent of Chikkim Formation in Kumaun is a formation of enormous thickness called Sangcha Malla Formation. This formation comprises wildeflysch and bears ophiolitic mélange with huge olistoliths of older ages, notably Triassic. The famous 'exotic blocks of Malla Johar are emplaced in this formation. The Formation comprises turbidites, graded bedded greywacke, chocolate coloured shale, radiolarian chert and foraminiferal ooze. The last contains well preserved Campanian forms namely *Globotruncanita calcarata, Globotruncana ventricosa, G. arca, Pseudotextularia elegans, Heterohelix striata* and others. This fauna is identical to that of Chikkim Formation of Spiti and Kangi La Formation of Zanskar. In the Kali–Gandaki valley of Nepal, the equivalents of Lapthal Formation, Spiti Formation and Giumal–Sangcha Malla Formations are called Lumachelle Formation, Saligram Formation and Kagbeni Group, respectively.

10.4.5 Indus Valley (Ladakh)

Indus Suture Zone constitutes a linear belt along which the Indian plate collided with the Asiatic plate and got subducted (see Chapter XIV for details). Along this suture the Tethys sedimentary sequence underthrusts Ladakh Granite batholith juxtaposing several tectonic units through a series of imbricate thrusts. The rocks involved in this suture zone range in age from Triassic to Eocene and the rock types range from flyschoidal sediments, carbonates, argillaceous rocks, volcanics, ultrabasic to intermediate plutonic rocks, ophiolites and ophiolitic mélange. While bulk of the Lamayuru Formation is of Triassic age (see Sec. 10.3.5), some portions especially those around Fotu La may extend up to Cretaceous. The presence of any Jurassic rocks in the Indus Suture zone is doubtful. However, the main Cretaceous sequence in this belt constitutes the basic volcanic suite, the Dras Formation. This formation extends from Burzil, through Minimarg, Dras and further onwards into Changthang. The lavas range in composition from basalt to spilite showing distinct pillow structure at many places, despite considerable deformation that they have suffered. Interbedded with the lavas are bands of limestone that bear various species of *Orbitolina*: namely, *O. kashmirica, O. discoidea, O. parma, O. pileus* and bivalve *Hippurites* indicative of Aptian to Cenomanian age for the volcanic rocks.

Large chaotically distributed blocks of similar *Orbitolina* bearing limestone together with volcanic rocks, volcanogenic sediments and serpentinites and several other rock types set in a matrix of deep-sea clastic sediments including radiolarian cherts constitute the ophiolitic mélange that extends for considerable distance from Pashkyum in the west to Khaltse in the east and is sometimes referred to as Khalsi Formation. Another band of similar mélange is known along the Shyok suture to the north. The ophiolitic melange is associated with rocks of blue-schist facies indicative of obduction from depth. Close together with the ophiolitic mélange (Khalsi Formation) is a flyschoidal unit namely Nindam Formation. The latter comprises graded bedded greywacke, turbidites, pelagic carbonates, chert, red and green mudstone and volcaniclastic sediments extending from Shergol to Hanle and may be of Late Cretaceous age probably passing upwards into Paleogene. Dras Formation constitutes the volcanism of island arc type. It is possible that Nindam Formation represents the forearc basin to the Dras volcanic arc that was generated in an intra-oceanic setting.

10.5 JURASSIC-CRETACEOUS COASTAL FACIES

10.5.1 Distribution

Jurassic and Cretaceous marine sediments are deposited in peninsular India due to frequent transgressions. As mentioned earlier, most of these transgressive phases are related to global eustatic variations. Coastal Jurassic sediments extend from Kutch in Gujarat to Barmer and Jaisalmer in Rajasthan. In Kutch, they are exposed in mainland and islands and hillocks of the Rann of Kutch. These hillocks are generally isolated by Quaternary cover that constitutes the main area of the Rann estuary. In Rajasthan, the outcrops are present wherever they are not covered by desert sand. Sub-surface drilling during oil exploration has revealed a continuous sequence in Jaisalmer basin and partial sequence in Barmer basin. Early Cretaceous rocks overlie the Jurassic rocks in Kutch as well as Rajasthan. However, more significant Cretaceous marine coastal sediments are exposed in southern India along eastern coast in Tamil Nadu (Vridhachalam and Ariyalur) and Puducherri. However, recent drillings for oil have revealed the presence of sub-surface Cretaceous sediments almost all along the eastern coast, notably in Cauvery basin and Krishna–Godavari basin. Cretaceous rocks have been encountered in offshore drilling in these areas as well. Fluvial and fluvio-marine deposition of Barremian–Aptian age took place on the east coast in Krishna Godavari and Palar basins. During Turonian, transgressive phase, an arm of the sea extended well into central India along Narbadha valley resulting in marine deposition that lasted for a limited period of time.

10.5.2 Kutch

The fossiliferous sequence studied for the first time in India happens to be the Jurassic sequence in Kutch. This is because of the highly fossiferous nature of the formation exposed in the mainland Kutch and the numerous ‘islands’ (hillocks during the retreat of the sea from the estuary) that dot the landscape of the Rann. The deposition in Kutch took place due to rifting during Jurassic times that continued through Cretaceous period. The rifting had been along the Precambrian orogenic trends of penisular India. Kutch rift was bounded by Nagar Parkar high in the north, Radhanpur–Barmer arch in the east and Kathiawar high in the south. The depression, thereby, became the site of coastal marine deposition during the transgressive and regressive cycles of Jurassic and Cretaceous. The sequence is punctuated by numerous breaks and diastems that reflect several transgressive and regressive sub-cycles from Middle Jurassic to Early Cretaceous.

The classification of the Jurassic–Cretaceous rocks of Kutch was attempted more than a century ago. The basic framework happens to be the same, despite numerous alterations and amendments made and refining of the fossil data from time to time. However, the classification in not in tune with the stratigraphic code that came into existence much later. In recent years, attempts have been made towards building a litho-stratigraphic classification; however, the same is complicated because of the lateral facies variation between the relatively deeper shelf sedimentation to the west and poorly fossiliferous shallow shelf facies with several breaks to the east. Thus, there are differences of opinion about the correlation of the various rock units. Accordingly, the original classification with some recent refinements is given in Table 10.3, although it is of an informal nature and is not in keeping with the code of stratigraphic nomenclature.

Table 10.3 Middle Jurassic-Early Cretaceous Sequence in Kutch

PERIOD	EPOCH	STAGE	FORMATION	SUB-DIVISIONS (informal) and LITHOLOGY	CHARACTERISTIC FOSSILS
			Bhuj	Sandstone, Shale	*Palmoxylon, Ptilophyllum*
		Albian		Ukra – calcareous Shale	*Australiceras, Colombiceras, Cheloniceras, Topaeum, Cleoniceras, Lemuroceras*
Cretaceous	Early	Aptian		Umia sandstone and shale	Unfossiliferous
		?Barremian	Umia (1000 m)		
		Valanginian		*Trigonia* beds	*Trigonia*
		?Berriasian		Barren sandstone	Unfossiliferous
				Umia ammonite bed	*Virgatosphictes, Micracanthoceras, Ptychophylloceras, Hemilytoceras, Umiatites*
Jurassic	Late			Upper Katrol shale	*Hildolochiceras, Dorsoplanites, Haploceras*
		Tithonian		Gajansar Beds	*Belemnopsis, Streblites, Phyloceras, Hildoglochiceras*

	Kimmerdigian	Katrol (300 m)	Upper Katrol sandstone	*Aulacosphinctoides, Virgatosphinctes*
			Middle Katrol (red sandstone)	*Waagenia, Katroliceras, Panchysphintes, Aspidoceras*
			Lower Katrol (sandstone, shale and marl)	*Lorquatisphinctes, Aspidoceras, Waagenia, Taramelliceras, Streblites, Ptychophylloceras, Hybnoticeras*
	Oxfordian		Kanthkot sandstone	*Epimayaites, Trigonia, Prograyiceras, Biplices, Torquatisphinctes, Prosophinctes, Ataxioceras*
		Chari (360 m)	Dhosa oolite (green and brown oolite)	*Taramelliceras, Mayaites, Perisphinctes, Discosphictes, Epimayaites, Paracenoceras, Peltoceratoides*
Middle	Callovian		Athleta beds (marl and gypseous shale)	*Peltoceras athleta, Orionoides*
			Anceps beds (marl and gypseous shale)	*Perisphinctes anceps, Indosphinctes, Reineckeia, Hubertoceras*
			Rehmanni beds (yellow limestone)	*Reineckia rehmanni, Sivajiceras, Idiocycloceras, Kellawaysites*
			Diadematus zone and Macrocephalus beds (shale with calcareous bands, golden oolite)	*Indocephalites diadematus, Macrocephalites macrocephalus, Dolicocephalites, Pleurocephalites, Kamptocephalites*
			Pachcham coral bed	*Macrocephalites, Stylina, Sivajiceras,Thamnastrea*
		Pach-cham	Pachcham shell limestone	*Macrocephalites, Trigonia, Corbula*
	Bathonian		Pachcham basal bed	*Corbula, Eomiodon*

While the sub-divisions are based mostly on ammonite fauna, in recent years a wide variety of other fossils, notably foraminifers, ostracodes, nannofossils and palynomorphs have been reported from these rocks. By the beginning of Albian, the regressive cycle resulted in the deposition of Upper Gondwana plant bed-namely, Bhuj Formation (Refer to Sec. 9.7.3).

10.5.3 Rajasthan

The Jurassic rocks in Rajasthan comprise Lathi Formation, Jaisalmer Formation, Baisakhi Formation and Bhadasar Formation whereas Cretaceous rocks comprise Pariwar, Sarnu and Habur Formations. Lathi Formation comprises arenaceous rocks mainly ill-sorted sandstone, purplish white clay and conglomerate. It bears Upper Gondwana flora, notably *Ptilophyllum acutifolium* and *Pterophyllum* sp. The deposition could have taken place in fluvio-deltaic conditions. Lathi Formation passes upwards with a gradational contact into Jaisalmer Formation. The latter is approximately 400 metres thick and comprises calcarenite and sandstone with an oolite bed towards the top. It bears ammonites which notably include *Indosphinctes, Reineckeia, Mayaites, Epimayaites* and foraminifers *Lenticulina, Trocholina.* The Formation can be correlated with Chari of Kutch and appears to be of Callovian–Oxfordian age. Baisakhi and Bhadasar Formations overlie Jaisalmer Formation and comprise black gypseous shale, oolite grits and sandstone, respectively. They bear Kimmeridgian fauna comparable to that of Katrol Formation of Kutch-namely *Lorquatisphinctes* and *Katroliceras.*

Bhadasar Formation of Jaisalmer basin is unconformably overlain by approximately 300-metres-thick Upper Gondwana plant-bearing formation comprising variegated silty sandstone of Early Cretaceous age named Pariwar Formation. The succeeding Habur Formation that has only limited extension is approximately 200 metres thick and comprises coquina limestone and calcarenite bearing *Deshayesites bodei* indicative of Aptian age. Sarnu Formation is possibly its equivalent or older and is restricted to Barmer basin.

10.5.4 South India

In the east coast area of south India, the Cretaceos marine deposition followed the rifting that produced the Bay of Bengal during Tithonian–Berriasian. However, while in Krishna Godavari basin the on-shore deposition commenced in Early Cretaceous times, in Vridhachalam, Ariyalur (Tiruchirapalli District) and Pondicherry (Puducheri) the oldest marine deposition commenced from Aptian onwards and continued up to the beginning of Cenozoic. These latter exposures have attracted the attention of the geologists from the middle of 19th century owing to their rich fossil wealth and easy accessibility. In recent years, the Cauvery basin and the entire east coast has been widely investigated for hydrocarbons and an enormous amount of sub-surface information has been generated about these rocks.

The marine Cretaceous rocks overlie the Upper Gondwana plant bed (Terrani plant bed) with a distinct unconformity marked by a band of conglomerate. The rocks dip at extremely low angles towards the sea side. The Cretaceous rocks are divided into three groups: Uttatur, Trichinopoly and Ariyalur. However, the Paleocene rocks that overlie them are grouped as Ninyur Group. All the groups have an unconformable contact with the overlying unit. They are further divided into formations as indicated in Table 10.4.

Table 10.4 Cretaceous Marine Sequence in Trichinopoly–Pondicherry Area

STAGE	GROUP	FORMATION (Thickness) Lithology	AMMONITE BIOZONES	PLANKTONIC FORAMINI-FERAL BIOZONES	OTHER IMPORTANT FOSSILS
Danian (Paleocene)		NINYUR (60 m) Limestone			*Nautilus danicus, Thalmannita madrugaensis*
Maastrich-tian	Ariyalur	KALLAMEDU (100m) Sandstone and clay			Dinosaurian fossil bones
		OTTAKOVIL (60m) Calcareous sandstone and siltstone	*Pachydiscus otokodensis*	*Globotruncana gansseri*	*Gunnarites kaklika, Stygmatopygus elatus*
		KALLAN-KURICHI (40 m) Limestone and marl	*Hauriceras rambda*	*G. stuarti*	*Pachydiscus egertoni, Exogyra tamulica*
Campanian		SILLAKUDI (500 m) Sandstone	*Karapadites karapadensis*	*G. ventricosa G. elevata*	*Inoceramus, Cretirhynchia plicatiloides*
Santonian	Trichi-nopoly	ANAIPADI (100 m) Sandstone and siltstone	*Placenticeras tamulicum – Kossmaticeras theoboldinum*	*Dicarinella asymetrica D. concavata Marginotruncana coronata*	*Arcotia indica, Damesites sugota, Pseudomaura nobilis*
Coniacian		KULAKKAL-NATTAM (50 m) Limestone, sandstone and siltstone	*P. tamulicum- K. recurrens*	*M. renzi*	*Mespuzosia indopacificum, Physa*
Turonian					
Cenomanian	Uttatur	KARAI SHALE/MARA VATTUR CLAY (400 m)	*Lewesiceras vaju Mammites conciliatum Eucalycoceras pentagonum*	*Praeglobotruncana helvetica Whitenella Rotalipora reicheli*	*Turrilites costatus, Scaphites kinge-anus, Rotalipora cushmani Lenticulina macrodisca, Ethelia alba* (algae)

Albian	KALLAKUDI LIMESTONE (75 m)	*Calycoceras newboldi* *Schloenbachia inflata*	*R. greenhornensis* *R .appenninica* *Planomalina buxtorfi*
Aptian			*Hedbergella planispira*
	GREY SHALE (50 m)		

The rocks are highly fossiferous bearing, ammonites, bivalves, gastropods, echinoids, crinoids, brachiopods, foraminifers, ostracodes, corals, shark teeth, algae, nannofossils and ichnofossils. The preservation of fossils is excellent in most of the formations. Some of the unusual elements include an 18-metre-long fossil conifer tree in Anaipadi Formation at Sattanur that has been preserved as a geological monument. Towards the top of the Ariyalur Group, the Kallamedu Formation bears numerous bones of dinosaurs namely *Megalosaurus, Titanosaurus* and *Bruhathakayosaurus matleyi* of Maastrichtian age.

The rocks were classified biostratigraphically on ammonite biozones. In recent years, hydrocarbon investigations on surface outcrops and bore holes have yielded better-controlled data; notably on the basis of planktonic foraminifers that have provided range zones, assemblage zones and interval zones. The correlation between the two kinds of biozones is indicated in Table 10.4.

10.5.5 Narmada Valley

The marine incursion along Narmada valley was restricted to a limited period towards the early part of Late Cretaceous, mainly Turonian, but also probably covering some portion of Cenomanian towards the lower part and that of Coniacian towards the upper. The rocks deposited comprise Bagh Group exposed as detached outcrops along the north bank of river Narmada from Barwah in Madhya Pradesh in the east to Rajpipla in Gujarat to the west. The maximum calculated thickness does not exceed 300 metres, though the total thickness is seldom exposed at any one place. The Bagh Group can be divided as follows:

Chirakhan Limestone (= Deola marl and coralline limestone)
Karondia Limestone (= Nodular limestone)
Nimar Sandstone

The lower part of Nimar sandstone is of fluviatile origin and is rapidly followed by estuarine conditions. The Bagh Group shows evidences of repeated transgressive and regressive phases and the first transgressive phase is towards the middle part of Nimar Sandstone when numerous ichnofossils appear in the sequence. These are followed by appearance of bivalves that continue throughout the sequence. The more common fossils include ammonites, *Placenticeras mintoi, Namodoceras scindia,* bivalves, *Inoceramus, Mytiloides, Astarte Birostrina,* gastropods, *Turritella, Ampulina*, echinoids, *Hemiaster*, calcareous algae and shark teeth. The fauna has a distinct Mediterranean aspect and is considerably different from that of Cauvery valley in south India. Bagh Group is overlain by Lameta Formation in Dhar region. Elsewhere, it is unconformably overlapped by Deccan volcanics.

■■■■■

11 Deccan Volcanics and Related Sediments

11.1 NOMENCLATURE AND DISTRIBUTION

The Deccan volcanic province is one of the largest flood basalt provinces of the world. It is the most prominent and widely distributed rock unit of peninsular India. It was given the name 'Deccan Trap' by W. H. Sykes in 1833. The word Deccan is derived from the Sanskrit word *dakshin* (meaning south) since the outcrops were located to the south of Vindyachal range. The rocks generate a profile of a series of step-like terraces because of which they were referred to as 'Trap'. The outcrops cover most of western and central India, including parts of states of Maharashtra, Karnataka, Andhra Pradesh, Telengana, Chhattisgarh, Madhya Pradesh, Gujarat and Rajasthan. While the area covered by the volcanics at present is over 500,000 square kilometers, the original extent of the lavas must have been much more widespread probably covering the terrain as far as Sindh and Baluchistan in the north west, Bundhelkhand in north-east, Jharkhand in the east and Rajamundry in the south-east, as is borne out by the remnants of detached outcrops and decomposed laterite derived from these rocks.

The detailed studies undertaken in recent years have helped in generating a lot of data on the distribution, composition and origin of this enormous suite of volcanic rocks. Geographically the lavas have been classified into three provinces namely northern province, western province and eastern province. The northern province covers the area to the north of the Narmada rift. The western province is by far the largest in distribution and thickness while the eastern province is the smallest covering the area of Satpura horst.

The lavas are thickest in western province and may be well over 2,000 metres in the neighbourhood of Mumbai. In Western Ghats, around Mahabaleshwar, the exposed thickness is over 1,300 metres and the total thickness may be much greater. Eastwards the thickness goes on gradually decreasing. In Sarguja in Chhatisgarh the thickness decreases to approximately 100 metres, whereas further eastward around Rajahmundry in Andhra Pradesh the thickness is approximately 60 metres. However, the original thickness must have been much greater since bulk of the cover seems to have been removed by erosion. It is obvious that the surface of the older rocks upon which the lavas extruded must have been extremely uneven and the topography determined the direction and the extension of the individual flows. The thickness of the lavas was evidently maximum in depressions, valleys and ravines. A remarkable feature of Deccan lavas is their persistent flatness and horizontal disposition. However, the flows are mildly tilted in some off-shore islands of Mumbai (Panvel flexure).

11.2 LITHOLOGY AND PETROLOGICAL CHARACTERS

The volcanic rocks have a uniform lithology comprising black or dark grey basalt associated occasionally with dolerite. The texture is variable ranging from dense glassy variety through fine-grained forms to porphyritic type. Most of the rocks are vesicular, the vesicles filled with varieties of zeolites. At many places, there are large amygdales filled with agate or variety of zeolites constituting the main part of the rock. Beds of volcanic ash associated with the lava flows are an extremely common feature.

While it is difficult to distinguish the individual flows in all cases, the thickness of each flow is variable from place to place. At Bhavnagar in Gujarat the average thickness of each flow is high at approximately 25 metres. Likewise, the flows are as thick as 30 metres in Chhattisgarh. On an average the thickness of each flow ranges from 12 to 30 metres, though thicker flows are also known. Individual flows extend for considerable distances indicating that the lava must have had a high degree of fluidity. Concretionary structures and columnar jointing are a common feature of most soft and hard rocks, respectively. The main primary minerals comprise labradorite, clino-pyroxenes and iron ore embedded in a mesh of interstitial glass. Olivine is rare and generally occurs in altered form as serpentine or iddingsite. A large variety of secondary minerals fill the amygdales, cracks and geode cavities. Some of the geodes may be as large as 60 to 70 centimetres. The most common secondary minerals filling these cavities are crystalline quartz, agate, jasper, bloodstone and opal. Less common minerals include a wide variety of zeolites, calcite, chlorophaeite and analcite.

Although Deccan lavas are uniformly basic in composition comprising primarily basalt and dolerite almost throughout their extension, there are some isolated localities where intermediate and acidic differentiates occur along with the basic volcanics. The most important localities that show this differentiation include Girnar complex, Osham hills and Pawagarh in Gujarat. Girnar igneous complex, that may probably represent the end phase of Deccan volcanics, contains a wide variety of plutonic and hypabyssal rocks ranging from intermediate to ultrabasic composition and alkaline rocks in addition to acidic, intermediate and basic lavas. Osham and Pawagarh hills display a thick cap of rhyolite in the central part showing flow structures. In Pawagarh there is a sequence of layered differentiates from basalt, through mugearite to rhyolite.

Bulk of the lava is of tholeiitic composition but it has been seen that the initial phase of volcanicity comprises alkaline magmatic activity as does the final stage.

The latter is clearly exemplified by the Girnar alkaline complex that contains widespread intrusives of alkaline rocks.

Dykes and dyke swarms of widely ranging composition are common in the western and northern Deccan provinces but they are rare in the central part. Most of these dykes are apparently feeders to the flows since they display a chemical and mineralogical similarity to them.

11.3 CAUSE OF DECCAN VOLCANISM

There are two differing opinions about the factors responsible for the generation of the magma of such gigantic dimensions. The earlier opinion that has again found favour in recent years is that pre-volcanic rifting along the west coast of India was responsible for injecting large amount of mantle derived melts that got extruded along fissures in the lithosphere. That the rifting along the west coast at the time of Deccan volcanic activity took place due to separation of Seychelles after that of Madgascar from India is undoubted, but whether that was responsible for the generation of magma is a matter of debate. The other opinion, generally finding more favour, is that during the course of northward drift of the Indian plate it moved over a hotspot which was responsible for the generation of magma from mantle material that erupted to the surface and continued till the plate left the hotspot leaving behind a trail of volcanic islands such as Chagos, Laccadive and Mascarene Plateau in the Indian Ocean. This theory, known as the plume theory has several adherents and is supported by data obtained from the mid-oceanic ridge in the Indian Ocean. However, whether magma generation is due to rifting or hotspot, what is undoubted is that mantle is involved with the magma generation in a big way. This is also borne out by the widespread distribution of xenoliths of mantle material throughout the Deccan volcanic province.

11.4 DECCAN VOLCANO-SEDIMENTARY SEQUENCES (INFRA- AND INTERTRAPPEAN BEDS)

11.4.1 Basement of Lava Flows

The Deccan Volcanics overlap a variety of rocks ranging from Precambrian peninsular basement to upper Gondwana and sediments like those of Bagh, Lameta and Umia formations to the north and marine infra-trappeans at Narsapur and Rajamundary. With most of these rocks they display an unconformable contact indicating that the lavas were extruded on the denuded surfaces of these rock units. The only sedimentary rock unit that has a distinct conformable contact with the basal lava flows especially in Jabalpur area is what is called Lameta Formation.

11.4.2 Infra-trappean Sedimentaries

The Lameta Formation, deriving its name from Lameta Ghat on river Narmada near Jabalpur, has a conformable contact with the basal lavas and can be conclusively classed as infra-trappean sedimentary unit. It rests conformably on Jabalpur Formation of Upper Gondwana group. In Jabalpur, the contact is marked by a characteristic green sand unit that constitutes the base of the formation. The main lithology comprises calcarenite and sandstone and the thickness varies from 20 to 50 metres. In Dhar region the Lametas are known to overlie marine Bagh Group.

The Lameta Formation is famous for the occurrence of the dinosaur remains. These were first described by Von Huene and Matley in 1933 from Bara Simla Hill section of Jabalpur Cantonment. The forms include various genera of sauropods notably *Titanosaurus* and several carnosaurs like *Indosuchus,* coelurosaurians like *Compsosuchus, Laevisuchus, Jubbalpuria* etc. and stegosaurian *Lametasaurus.* However, recent researches on the Lameta dinosaur fauna from Jabalpur, Pisdura and Dongargaon in Central India, and Rahioli in Western India have recognised only two genera of tiatanosaurid dinosaurs (*Jainosaurus* and *Isisaurus*) and three genera of abelisaurid (carnivores) namely *Indosuchus, Indosaurus* and *Rajasaurus.* In recent years, a large number of dinosaur egg shells have been discovered from widely separated localities of infra-trappean and intertrappean rocks including those from Lameta Formation. In addition to isolated egg shell fragments, a large number of nesting sites with dinosaur egg clutches extending across western and central India have also been documented. All the nesting sites occur in hard sandy carbonate rocks of the Lameta Formation, suggesting strong nesting site selectivity. Apart from dinosaurs, the other fossils found in Lameta Formation include fishes: *Igdabatis, Rhombodus, Lepidosteus, Pycnodus, Eoserranus, Osteoglossiformes, Siluriformes,* turtles: *Shweboemys pisurensis,* Boidae snakes, Dyrosauridae crocodiles, mollusks: *Physa prinsepii, Paludina normalis* and *Lymnaeasubulata* and a wide variety of ostracodes and charophytes.

Apart from the Lameta Formation, infra-trappean marine sediments have been identified in subsurface wells drilled for oil in Krishna-Godavari basin. These sediments are rich in planktonic and benthic foraminifers and dinoflagellates.

11.4.3 Intertrappean Beds

The thin sedimentary beds interbedded with the lavas, referred to as inter-trappean beds, are widely distributed from Kutch in the west to Mandla in the east, Lalitpur in the north to Gurmatkal in the south and Rajahmundry in the south-east. Most of these are rarely more than a few metres in thickness and are often less than a metre. They occur in long broken outcrops that seldom continue beyond a kilometre, wherever they can be traced laterally. They are generally similar to Lameta Formation in lithology, but Lameta Formation comprises predominantly limestone-sandstone sequence, whereas limestone is less common and sandstone, siltstone, mudstone, clay, carbonaceous shale and chert are the most common lithologies of the intertrappean beds.

The inter-trappean sediments were earlier thought to be deposited only in fluvio-lacustrine basins. However, subsurface studies undertaken in Narsapur, Elamanchili, Duddukuru, Modi and other places have shown that the deposition of these beds has taken place in shallow marine conditions. They contain a rich foraminiferal assemblage. The inter-trappean beds at Asifabad and Chhindwara area indicate deposition in brackish water conditions. From this, it is clear that depositional environments were not uniform throughout the length and breadth of Deccan Trap province, though bulk of the deposition, especially towards the central part, was of a fluvio-lacustrine type.

The intertrappean rocks are highly fossiliferous but the fossils are in no way different from those of the infra-trappean beds indicating thereby that the life in the area was not much affected by the environmental crisis induced through volcanism, at least in the early phase of volcanic activity. The beds deposited under fresh water conditions contain a variety of organisms comprising fishes, amphibians, reptiles, mammals, gastropods, bivalves, charophytes and a variety of palynological assemblages. Dinosaur teeth were reported by Lydekker in 1890 from intertrappean beds but in recent years theropod teeth, bones and egg-shell fragments were recovered from widely separated localities including Anjar in Kutch, Nagpur in Maharashtra and Asifabad in Andhra Pradesh. Teeth of Eutherian mammals have also been found from the intertrappean beds of Naskal in Andhra Pradesh. The marine intertrappean outcrops of Rajahmundry and the subsurface sections from the adjoining areas of Krishna-Godavari basin contain foraminifers and dinoflagellates.

11.5 AGE OF DECCAN VOLCANICITY

The age of Deccan volcanism remained a subject of debate for many decades during the twentieth century. The ages proposed from time to time ranged from Late Cretaceous (Turonian) to Oligocene. In fact, earlier the volcanicity was thought to have extended for a considerable length of time covering nearly 50 million years. This was on the basis of interpretation of the mollusks, fishes and plant fossils occurring in the intertrappean beds. Radiometric dates obtained through K/Ar method also resulted in erroneous interpretations showing a wide dispersion in ages due to argon loss and some other primary causes. However, in recent years following detailed investigations, we have a clear perception of the relatively very short time span of the total volcanic episode.

The earlier opinion that the volcanism commenced in post-Turonian times (possibly Santonian) was supposed to be based on the age of Bagh group of rocks that were thought to be lateral facies variants of Lameta Formation. However, Lameta Formation is known to overlie Bagh beds in Dhar region and is decidedly younger. This is also borne out by the fauna and flora obtained from this formation and other infra-trappean equivalents. Likewise, the assignment of upper age limit of Early Tertiary (even Oligocene), supposedly based on fishes (*Lepidosteus, Eoserranus* etc.), plant assemblages (notably *Azolla*) and charophytes obtained from intertrappean beds has also been found to be erroneous.

The interpretation based on the earlier reported fauna and flora and recent findings all indicate that a major part of Deccan volcanism was limited to a minor time span and restricted to Late Maastrichtian age. *Azolla cretacea* is a typically Late Cretaceous form while most of the other cited forms in earlier literature are unreliable for establishing age relationships. Of particular significance, as far as age is concerned, are the dinosaurs that were thought to have become extinct during Deccan volcanism. However, the recent findings have shown widespread distribution of dinosaurs and dinosaur eggs and nesting sites in the infra-trappeans and dinosaur bones, egg shells and teeth in the inter-trappean beds, indicating that dinosaurs had survived the early phases of Deccan volcanic episode. However, the upper levels of Deccan Traps, particularly in the western Ghat region, Lalitpur in Bundelkhand region, and Rajahmundry in the Krishna-Godavari basin extend into Early Paleocene (62 m.y). The recent report of fishes from infra-trappean and inter-trappean beds, namely *Iqdabatis, Rhombodus, Lepidotes, Lepisosteus* suggest a Late Cretaceous age. Likewise, ostracodes *Paracypretta, Mongolianella, Condoniella, Cypridopsis* and *Cytheridella* and *Aquilapollenites-Ariadnaesporites* palyno-floral assemblage reported recently from many intertrappean outcrops all support a Maastrichtian age for these beds. The marine fauna from the infra-trappean beds obtained from sub-surface well sections in Krishna-Godavari basin have also yielded foraminifer species namely *Abathomphalus mayaroensis,* and the lower inter-trappean beds have yielded *Globotruncana stuarti*, both of typically Maastrichtian age. The upper intertrappean beds and the supratrappean beds overlying the topmost lava flows of this area yielded Early Paleocene foramninifers.

Radiometric dates obtained from lava flows at various levels have given variable results, especially when they have been obtained through K/Ar method. However, more accurate dates have been obtained through Ar/Ar method and they fall within the range of 62–67 million years. This range accords well with the evidence of fossils from the subsurface sections of Krishna Godavari basin, as mentioned above. Further it has been suggested that 80% of Deccan traps erupted over a very short period of 1 m.y. close to Cretaceous-Tertiary boundary (KTB)

Perhaps one of the most conclusive evidence that supports a brief span of volcanic activity restricted to less than 5 million years within Maastrichtian age is that obtained from palaeomagnetic studies. Based on extensive palaeomagnetic studies carried out in western and central Deccan provinces in recent years it is seen that northern and central areas show a reverse magnetic polarity whereas the lavas at Mahabaleswar show a normal polarity. Bulk of the lava (according to some workers 85%) shows a reversed polarity. This evidently corresponds to C29r polarity chron and possibly a part of C30n chron (see para 1.7). Very few sections show r-n-r sequence that would seem to cover C30 chron and C29r sub-chron. Because of this limited time interval many workers believe that the total duration of bulk of Deccan volcanism may not have exceeded 1 million years.

11.6 CRETACEOUS-TERTIARY BOUNDARY

11.6.1 Global concept of Cretaceous-TertiaryBoundary (KTB)

A major extinction event in Phanerozoic times has been at the end of Cretaceous marking its boundary with Tertiary. The percentage of the species that became suddenly extinct during this event is more than that of any other event, except for the Permian-Triassic boundary event. However, the extinction was much more spectacular than in the latter case, since it involved the elimination of a major group of land animals that dominated the stage throughout Mesozoic, the dinosaurs, and another group ruling the marine environment, the ammonites. Apart from these a host of faunal and floral species met their doom during this event that transformed the scenario totally at the commencement of Cenozoic era.

One of the major problems that confront all earth scientists is to locate the cause of such a catastrophic event. For many years, various theories were proposed but none of these were found to be sufficient to explain this mass extinction event. An important geochemical factor that marks the event at the termination of Cretaceous period is a sudden increase in the level of the trace element iridium. This is known as the KTB iridium anomaly. Many marine and fresh water successions throughout the globe exhibit a high peak of iridium at a particular level that marks the boundary. Iridium is extremely rare on the Earth's surface but is known from the mantle or meteorites. Since the boundary is marked by this anomaly on a global scale, it is necessary to find the source of the high quantities of iridium. Two theories are in vogue to explain this anomaly and thereby the cause of extinction.

Alvarez (father and son) and their group in 1980 proposed an extra-terrestrial cause for the mass extinction at the termination of Cretaceous. It was proposed that the Earth was impacted by an extra-terrestrial object, an asteroid or a meteorite, of enormous dimensions which generated a cloud cover that dislocated the ecosystem of the Earth's surface resulting in climatic disturbances and break-up of the food chain leading to mass extinctions. The iridium anomaly was due to the large quantity of iridium brought in by the impacting object from outer space. The evidences of the impact site are reported to have been located at Chicxulub in Yucatan Province of Mexico.

The other hypothesis proposes that Deccan volcanic activity was of such enormous proportions that it introduced physical and geochemical changes in the global environment. This included the introduction of volcanic dust in the atmosphere obscuring the sun and acid rain precipitation that damaged the environmental balance. Some authors calculated that billions of kilograms of CO_2 and SO_2 were released into atmosphere during the course of this volcanic episode. This could have brought about a global transformation in the environment destroying the ozone layer and bringing about climatic changes that were responsible for mass extinction of some dominating groups of organisms that had reached a high level of specialisation. It is argued that iridium could have been derived from upper mantle that was involved in the volcanic activity, since it is known that deep seated shield volcanoes emit iridium derived from upper mantle.

Comparing the evidences in favour or against each of these theories, the extra-terrestrial cause appears to have more points in its support. Accordingly, most earth scientists believe that the impact of an extra-terrestrial object could have triggered the environmental changes that brought about the extinction event. One of the special features of the extinction event is that by and large it has been of a stepped type but the time interval between the various levels of extinction is small and more or less restricted to uppermost part of Maastrichtian. Marine regressions and Deccan volcanism over an extended period at the end of Cretaceous may have significantly contributed to the environmental stress and ultimate disruption of stressed ecosystems may have been caused by the asteroid impact. Accordingly, the K/T boundary is fixed between Maastrichtian and Danian dated to 65.5 ± 0.3 million years

11.6.2 K/T Boundary in India

For locating a stratigraphic boundary, it is necessary to have a continuous section preferably in a marine sequence. In India, most of the sequences in Maastrichtian are interspersed with volcanic rocks, leaving only the inter-trappean rocks for such investigation. In most of these sequences the iridium anomaly is not clear. The marine sequence where iridium enrichment has been observed at the boundary level happens to be in Meghalaya. Here the uppermost Cretaceous rocks constitute what is known as Mahadeo Formation. The boundary on the basis of the extinction of *Globotruncana stuarti* is approximately 10 metres below the contact between Mahadeo and overlying Langpur Formation. In non-marine sequence three levels of iridium enrichment have been reported in the inter trappeans of Anjar in Kutch. Dinosaur egg shells and Maastrichtian ostracodes occur in between and above these levels. Owing to the diffused nature of the iridium-bearing limonitic veins and occurrence of Maastrichtian fossils above the topmost iridium level, the iridium levels of Anjar intertrappean section were interpreted as leached products of Deccan volcanic flows.

In subsurface wells drilled for hydrocarbons in Krishna–Godavari basin, as mentioned above (Sec. 11.5), Maastrichtian microfauna—namely, planktonic foraminifers and dinoflagellates — occur below the lavas as well as within the lower inter-trappean beds, while Paleocene (Danian) fauna is seen in the upper intertrappean, and overlying supratrappean beds. Early Paleocene foraminifers have also been reported from the intertrappean outcrops of Rajahmundry. Evidently, the K/T boundary lies somewhere near the end of the volcanic activity in the upper intertrappean beds of the subsurface sections.

12 Cenozoic Sedimentation

12.1DISTRIBUTION OF CENOZOIC ROCKS

Early Cenozoic rocks (Paleogene and Early Neogene) are widely distributed as marginal deposits almost all around the Indian craton. They display rapid facies changes both laterally and vertically, ranging from marine through estuarine to fluvatile conditions of deposition.

Towards the north-western side, the Indian plate was rapidly closing in for its rendezvous with the Asiatic plate by the end of Cretaceous. The collision between the two plates took place by the beginning of Cenozoic (for details see Chapter 14). This brought an end to the Tethys basinal sedimentation leaving behind some remnants along the northern margin of Zanskar basin, Indus and Shyok suture zones where Early Cenozoic marine sediments were deposited. These were followed by intra-montane fresh water deposits that persisted through Late Neogene and Early Paleogene. Pleistocene and Holocene lacustrine and glacial deposits are present as isolated outcrops all along Higher Himalaya.

Following the collision and consequent subduction and initial phases of Himalayan uplift, a basin developed in front of the newly emerged orogen along what is presently the Lesser and Outer Himalayan zone. This became the site of sedimentation with the sediments derived from areas of positive relief to the north and possibly from the south. As the uplift continued, the basin underwent rapid changes resulting in the retreat of the sea and shifting of sedimentation from marine to paralic, estuarine and eventually fluviatile deposition. The retreat started from east to west due to which the marine sedimentation continued well up to beginning of Miocene in Sind-Baluchistan and Kutch, while it ended in Eocene in Jammu and Himachal Pradesh. The Siwalik Group, constituting the fluviatle sediments was deposited as a continuous belt from east to west as the frontal molasse facies (see Chapter 13).

In north-eastern India, a thick pile of Paleogene and Neogene sediments was deposited following the collision between India and Myanmar (Burma). Early Paleogene rocks comprised marine rocks that were followed by fresh water sediments. There is a facies difference between rocks deposited in Upper Assam, Surma valley, Khasi and Jaintia hills. In north-eastern India, also like that of Sind and Baluchistan, the sea receded much later than in the central parts of Himalayan foothills.

The marine Paleogene arm of the sea that existed in north-western part of Lesser Himalaya extended into south up to Rajasthan and Kutch. However, the sediments are covered by Indo-Gangetic alluvium and desert sand of Rajasthan exposing only a few inliers in the Bikaner–Nagaur region, Jaisalmer and Barmer.

Along the western coast in Kutch marine Paleogene-Early Neogene sediments overlying Deccan Trap are exposed. These sediments are only partially exposed in mainland Gujarat but are present in subsurface area. The drilling for hydrocarbons in on shore and off shore areas in western India has indicated a thick succession of marine shelf sediments overlying Deccan Trap covering major part of Paleogene rocks.

In South India, the Cretaceous succession in Tiruchirapalli region is unconformably overlain by marine Paleocene Niniyur Group (see Table 10.4). Neogene sediments in Tamil Nadu comprise fluviatile deposits namely Cuddalore Sandstone that provides a sedimentary cover dissected by present-day streams resulting in occurrence as isolated outcrops. The corresponding rocks in Kerala constitute Quilon Formation and Warkalli Formation.

The best and nearly continuous exposures of marine Cenozoic rocks occur in the group of Andaman-Nicobar Islands.

12.2 HIGHER HIMALAYAN PALEOGENE INTRA-MONTANE BASINS

12.2.1 Marine Sedimentary Sequence

Towards the northern margin of Zanskar basin, bordering the Indus suture zone, the uppermost Maastritchtian rocks are overlain by Paleocene and Early Eocene nummulitic limestone (refer to Sec. 10.4.3). This sequence (variously named as Kesi Fm. constituting the outer shelf deposition and Dibling Fm. the equivalent inner shelf deposition) bears a rich foramiferal fauna dominated by rotaliids namely *Lokhartia heimei, Rotalia* cf. *trochidiformis* associated with other smaller foraminifera towards the base. Higher up, the sequence bears other foraminifers namely *Daviesina danieli, D. khatiyali, Ranikothalia* sp., *Fasciolites, Orbitolites.* The fauna indicates an age range from Paleocene to earliest Eocene (Ypresian). These formations have a special significance since they indicate the final closure of Tethys and help in fixing the time of collision between the Indian plate and Asiatic plate (see Chapter 14 for details). The equivalents of these rocks in the Indus fore-arc basin belonging to a flyschoidal facies, are represented by the uppermost part of Nindam Formation that started depositing in late Albian and continued upto earliest Eocene (see Sec. 10.4.5).

12.2.2 Intra-montane Fresh Water Sequence

After collision followed by subduction-obduction, an intramontane basin was formed along the Indus suture zone that became the site of fluviatile and lacustrine deposition. This group of sediments popularly referred to as Ladakh or Kargil molasse, constitutes the Indus Group and consists of three formations namely Kargil,

Tharmusa and Pashkyum that have a vertical and lateral contact relationship and could be time transgressive. The group has a total thickness of approximately 1800 m and consists of conglomerate at the base and top with intervening thick beds of massive sandstone and interbedded purple mudstone. Unlike the older flyschoidal rocks namely Nindam and Lamayuru Formations that are exposed as separate tectonic units bound by thrusts, the Indus Group is autochtonous and overlying unconformably the Ladakh granodiorite massif. The latter was emplaced in post Cretaceous times since it cuts the Dras Formation (see Chapter 14). The environment of deposition appears to be that of alluvial fans at the base followed by lacustrine and fluvial sedimentation. Fossils of fresh water mollusks, fish scales and vertebrates notably *Hyboops* have been recovered from the sequence. The fossils are indicative of an age range from latest Eocene to Oligocene. In lithology and other characters the sequence resembles, to a large extent, the Murree Formation (see Sec. 12.3.2). The equivalents elsewhere in Tibet could be the Kailash Conglomerate.

12.3 WESTERN HIMALAYAN FRONTAL ZONE

12.3.1 Marine Paleogene Sequence

While the main Tethys sequence culminated by the beginning of Cenozoic along the northern margin bordering Indus suture line, a new marine basin emerged to the south of the Higher Himalayan zone. This basin constituted a narrow marine seaway in the Lesser Himalayan zone that extended from Sind-Baluchistan in the west, through Salt Range, Jammu Hills, Bilaspur-Simla region of Himachal Pradesh, Garhwal and extending into Nepal. The same seaway extended to Bikaner and Jaisalmer as indicated by sub-surface deposits and was a part of Early Cenozoic basin of Kutch. While the marine deposition in Sind-Baluchistan and coastal region of Kutch extended for the entire range of Paleogene, in Jammu hills, Himachal and Garhwal it was restricted from Late Paleocene to Middle Eocene. The sea retreated gradually westwards due to which no marine sediments younger than Middle Eocene (Lutetian) were deposited in these areas.

The marine Paleocene–Eocene sequence rests on a Precambrian basement. In Dogadda area of Garhwal it rests on Tal Formation, in Himachal on Tundapathar Limestone, Shali, Blaini and Simla Slate and in Jammu hills on Sirban Limestone. The formation is thickest in Himachal where it ranges over 700 metres. In Garhwal it is only 170 metres, while in Jammu hills it is approximately 150 metres. In Himachal, while part of it occurs as isolated outliers comprising pisolitic laterite/bauxite at the base followed by quartz arenite, variegated shale and limestone (Kakara Formation), the main outcrop shows a complete thickness and is referred to as Subathu Formation. It passes conformably upwards into younger paralic and fresh water sediments (Dagshai and Kasauli Formations; see Sec. 12.3.2). In Jammu hills also the bottommost units constitute pisolitic laterite/bauxite that pass upwards into lithology identical to that of main Subathu Formation. In Poonch region in western part of Jammu hills, thin band of marine Eocene rocks is exposed within the schuppen zone of imbricate thrusts overlying Panjal Volcanics and equivalents of Zewan Formation.

The basal Subathu constitutes black-grey facies characterised by carbonaceous shale that in Jammu area contains coal deposits. These beds are intercalated with green sandstone. The next unit constitutes the green shale facies comprising olive green splintery shale interbedded with thin limestone and overlain by massive limestone bearing occasional bands of coquina composed of oyster shells. The topmost unit constitutes red bed facies consisting of red shale capped by white quartzite that marks the transition to Dagshai/ Murree formation. The last facies marks transition to brackish water deposition.

Subathu Formation bears a rich and varied fauna and palynoflora. The basal unit in the type area (Subathu in Himachal) bears a large marine mammal *Ishatherium subathuensis* associated with shark and ray teeth. Other fauna includes a wide variety of mollusks, ostracodes. The distinctive fauna includes age specific foraminifers. The foraminifers towards the lower part of the sequence include *Lokhartia heimei, L. conditi, L. tipperi, Rotalia bensoni, R. trochidiformis* indicative of Paleocene age. The upper part of the sequence bears *Nummulites atacicus, N. mamilla, Assilina granulosa, A. davies* of Early Eocene (Ypresian) age followed by *A. spira* indicative of Middle Eocene age. While bulk of the fauna appears to be of Ypresian age, the total range is from Late Paleocene to Middle Eocene (Thanetian to Lutetian). In the uppermost part of the sequence namely the red bed facies, a vertebrate bone bed has been located in Kalakot, Metka and Mehogla outcrops of Jammu hills. Such a bone bed is also known to occur in Pakistan at the same stratigraphic level. This bone bed comprises remains of tapiroids, rhinoceratoids and brontotheres. This bone bed marks the initiation of non-marine sedimentation in Lesser Himalaya.

12.3.2 Fresh water Paleogene-Early Neogene Sequence

The marine Eocene rocks are conformably overlain by fresh water deposits both in Simla and Jammu hills. The transition is marked by the deposition of red bed facies that shows brackish water environment, topped by quartzite that could indicate deposition of a blanket of coastal sand as the sea eventually withdrew and was replaced by deltaic and fluviatle sediments. The Late Eocene–Oligocene–Early Miocene rocks are variously named. In Jammu hills, they constitute an extension of the rocks widely exposed in Pakistan, referred to as Murree Formation. In Kangra region of Himachal Pradesh they are referred to as Dhramshala Formation while in Simla hills they constitute the Dagshai and Kasauli Formations that together with the underlying Subathu Formation are sometimes grouped as Sirmur Group.

The Murree Formation of Pakistan is an eastern equivalent of marine Gaj Formation of Sindh. It extends around Jhelum syntaxis into Poonch and Jammu hills. While the formation has an enormous outcrop width in Poonch, the same decreases considerably eastwards in Batote- Sudmahadev area. The large outcrop width to the west is a result of repeated isoclinal folding. The total thickness of Murree Formation in Jammu region does not exceed 2,000 metres. The Murree Formation consists of hard fine-grained sandstone with bands of pseudo-conglomerate and red and purple splintery shale. The upper part of Murree Formation closely corresponds in lithology with the overlying Siwalik rocks into which it passes upwards imperceptibly. The fauna and flora is scanty and restricted to shale bands. It is known to bear some vertebrate remains notably those of *Crocodilus* and *Dinotherium.* In addition, there are bands bearing fresh water mollusks and monocotyledonous and dicotyledonous leaf impressions. Most of these forms are not age conclusive. There is a transitional passage in Kalakot and elsewhere between Subathu Formation and Murree Formation and therefore the earlier opinion that Murree Formation represents only Early Miocene age and Oligocene is absent in Jammu does not seem to be valid. In all probability, on the basis of underlying Subathu and overlying Siwalik rocks, the age range of Murree Formation could be from Late Eocene to Early Miocene.

As in Jammu hills, the Dagshai Formation overlies the Subathu formation with a gradational contact and is followed likewise by Kasauli Formation. The basal beds of Dagshai Formation consist of pisolitic marl. The main lithology comprises current bedded sandstone alternating with purple shale. There are also occasional bands of conglomerate. Kasauli Formation is essentially a sandstone formation with subordinate shale beds. The difference from Dagshai is in the infrequency of shaly beds. As in the case of Murree Formation there is a conformable transition of Kasauli Formation into overlying Siwalik Group. Fossils include palm leaves and fresh water mollusk shells. The age range of Dagshai and Kasauli Formations appears to be the same as that of Murree Formation. (For details refer to Table 12.1).

Table 12.1 Paleogene Stratigraphy of North-Western Lesser Himalaya

Jammu	**Himachal**	**Uttara-khand**	**General lithology**	**Distinctive fossils**	**Facies**	**Age**
Rock Formation	Rock Formation	Rock Formation				
(Max. Thick-ness)	(Max. Thick-ness)	(Max. Thick-ness)				
Murree Formation (2000 m,)	Kasauli Formation		Sandstone, purple shale, conglomerate	Palm leaves, fresh water mollusks	Fresh water	Early Miocene
	Dagshai Formation		Pisolitic marl	*Crocodilus, Dinotherium*	Paralic Estuar-ine	Oligocene
Subathu Formation (150 m.)	Subathu Formation (700 m.)	Subathu Format-ion (170 m.)	Carbonaceous shale, green and olive shale, thin limestone	Vertebrate bones *Assilina spira Nummulites atacicus, N. mamilla, Assilina granulose, A. davies*	Marine	EOCENE Lutetian Ypresian
				Lockhartia heimei, L. conditi, L. tipperi, Rotalia bensoni, R. trochidiformis	Marine	PALEOCENE Thanetian
	Kakara Formation		Pisolitic laterite, variegated shale			

Murree Formation and Kasauli Formation are conformably overlain by Siwalik Group of rocks that extends throughout the foothill zone of Himalaya and ranges in age from Middle Miocene to Early Pleistocene. (For the details of Siwalik Group refer to Chapter 13).

12.4 EASTERN HIMALAYAN FRONTAL ZONE

The north-eastern region of India got squeezed after Middle Cretaceous by the closing in of Indian plate. Due to subduction that followed, varying facies of contemporaneous deposition evolved during Paleogene, depending on the tectonic nature of the basin. Closer to the subducting Indian plate shelf facies was deposited replaced by slope sediments eastwards. The marine conditions persisted in this region almost upto the beginning of Miocene like that of the western region of Sindh-Baluchistan. However, the lateral facies changes and paucity of fossil material impede the process of time correlation, due to which corresponding lithological units have been given different names when they occur in a different setting. Because of the squeeze, a schuppen zone evolved along the border with Myanmar in which various rock units got involved complicating the stratigraphic resolution. The broad stratigraphic succession is given in Table 12.2.

The Late Cretaceous–Cenozoic sequence rests on Meghalaya plateau towards its southern side. The contact between Cretaceous rocks including Sylhet Trap and Cenozoic rocks is marked by conglomerate indicative of a break in deposition. However, in Um Sohryngkew river section in Therria area there appears to be a continuous sequence whereby the Cretaceous Mahadeo Formation passes conformably upwards into Langpur Formation. This section is probably the only one where Cretaceous–Tertiarty boundary has been marked in a marine sequence in India (see Sec. 11.6.2).

In the schuppen zone along India-Myanmar border, a completely different facies of rocks comprising mainly of splintery grey shale with few bands of sandstone is widely exposed in Nagaland and Manipur. This formation named as Disang Group represents the outermost shelf and slope facies and is in all probability the eastern equivalent of Eocene rocks namely Sylhet and Kopili Formations.

In Lower Assam, the Barail Group is one of the thickest units and has a widespread distribution. This Group has an economic importance as it bears coal measures and together with Tipam Sandstone is a repository of hydrocarbons. There is an unconformable contact between Barail and overlying Surma Group, marked by a persistent band of conglomerate both in Upper and Lower Assam.

Table 12.2 Cenozoic stratigraphy of north-eastern India (Meghalaya, Assam, Nagaland, Arunachal)

Epoch	Group	Formation (Maximum Thickness in metres)	Lithology	Characteristic Fauna/flora	Equivalents	Facies
Pliestocene	Dupitila				Dihing Namsang	Fresh water
Pliocene		Girujan Clay (2,300)	Soft mott-led clay with bands of sandstone			Fresh water, fluvial
Miocene	Tipam	Nahorkatiya Sandstone (=Tipam Sandstone) (2,000)	Massive sandstone with thin bands of shale	Vertebrate: *Anthracotherium* Palynomorphs: *Acanthotrilites, Quercoidites*		
	Surma	Boka Bil (1,500)	Shale with sandstone bands	Mollusc shells, Palynomorph *Deltoidospora*		Estuarine, Coastal marine
		Bhuban (2,400)	Alternating sandstone &mudstone	Mollusc shells, *Cidaris* sp.		Shallow marine
Oligocene	Barail	Renji (1,000)	Massive sandstone	Foraminifera: *Haplophragmides, Miliammina, Trochammina* Palynomorphs: *Meyeripollis noharkotensis, Longapertites* sp.		Marine
		Jenam (1,200)	Shale with bands of sandstone			Brackish water
		Laisong (2,400)	Sandstone with bands of shale			Estuarine
Eocene	Jaintia	Kopili (450)	Varying lithology with sandstone and shale	Foraminifera: *Nummulites beaumonti, Assilina spira*	Disang	Marine
		Sylhet (500)	Limestone	Mollusc shells	Prang Umlatdoh Lakadong Therria Langpur	Marine
Paleocene		Tura (=Cherra) (200)	Sandstone			Fresh water
Cretaceous					Mahadeo	Marine

12.5 PALEOGENE-EARLY NEOGENE SEDIMENTATION IN WESTERN BASINS

12.5.1 Rajasthan

The same Paleogene seaway that covered Sindh-Baluchistan region and extended as far as Garhwal in the east (see Sec. 12.3.1) continued into Rajasthan and Kutch region, the intervening area being covered by Indo-Gangetic alluvium and desert sand. As in the case of Jammu hills and Himachal Pradesh, the seaway was restricted in time to Paleocene to Middle Eocene and retreated thereafter. The outcrops in Rajasthan are confined to three basins from north to south namely Bikaner–Nagaur basin, Jaisalmer basin and Barmer basin. Of these maximum exposures are present in Jaisalmer basin. The lithostratigraphic nomenclature in three basins is different for equivalent formations.

In the Bikaner–Nagaur basin, the Paleocene–Eocene succession consists of Palana, Marh and Jogira Formations. The sequence is fossiliferous, especially Jogira Formation that contains nummulitic limestone bearing *Nummulites atacicus* and *Assilina granulosa* indicative of Early Eocene age. It also bears lignite seams and bands of fuller's earth. Most of the sequence is covered by desert sand and can be approached only by shallow drilling or trenching.

Paleocene–Middle Eocene sequence of Jaisalmer basin is represented by Sanu, Khuiala and Bandah Formations. Sanu Formation comprising sandstone-shale alternations unconformably overlies Pariwar/Habur Formations of Early Cretaceous age (see Sec. 10.5.3). It bears glauconitic clay at the base. The fossils include gastropod shells and fish teeth. The unconformably overlying Khuiala is highly fossiliferous bearing bands of nummulitic limestone and is overlain conformably by bioclastic limestone of Bandah Formation.

Barmer basin sequence is represented by Fatehgarh, Akli, Mathaji Ka Dungar and Kapurdi Formations. Of these Akli and Kapurdi Formations are highly fossiliferous and the age indicated for the sequence ranges from Paleocene to Early Eocene.

12.5.2 Kutch

Cenozoic rocks in Kutch display a marine sequence from Paleocene to Pliocene. In that respect, this is one of the few areas in India where such a fossiliferous sequence is exposed. The Paleocene comprising Matanumadh Formation which is equivalent of Marh Formation (see Sec. 12.5.1) is poorly fossiliferous. It comprises variegated clay, tuffaceous shale and sandstone. Early and Middle Eocene comprising Naredi Formation, Harudi Formation and Fulra Limestone is highly fossiliferous and bears various species of planktonic foraminifera like *Globorotalia, Globigerina* and benthic genera *Gypsina, Assilina and Nummulites.* There is no deposition in Late Eocene and Middle Eocene is directly overlain by Oligocene rocks namely Maniyara Fort Formation that bears rotalids and miliolids and several other benthic forms and is indicative of the shallowing of the sea. There is a break towards the end of Oligocene but Miocene marks a transgression whereby the sea seems to have encroached deep into the mainland Kutch where the Miocene rocks directly overlap the Deccan Traps. The Miocene rocks constituting Khari Nadi Formation bear a wide variety of foraminifers, notably species of *Miogypsinoides, Lepidocyclina, Ammonia, Gypsina* and several miliolids indicative of an Aquitanian/Burdigalian age. The overlying Kankawati 'series' that is locally exposed bears long ranging miliolids and ostracodes and could be of Pliocene age.

Table 12.3 Marine Cenozoic Stratigraphy of Kutch

Epoch	Age	Formation	Lithology	Characteristic Fauna/flora
Miocene	Aquitanian/ Burdigalian	Khari Nadi	Coarse sandstone	*Miogypsinoides dehaarti*
Oligocene	Chatian			
	Rupelian	Maniara Fort	Yellowish siltstone, massive limestone	*Nummulites fichteli, N. intermedius*
Eocene	Priabonian			
	Bartonian	Fulra Limestone	Foraminiferal limestone	*Truncorotaloides rohri, Catapsidrax unicavus, Orbulinoides beckmanii*
	Lutetian	Harudi	Grey shale, Calcareous claystone	*Globigerina frontosa, Globigerapsis kugleri, Hantkenina aragonensis*
	Ypresian	Naredi	Claystone, dark shale, limestone with gypsum	*Assilina granulosa, Nummulites atacicus*
Paleocene	? Thanetian	Matanumadh	Variegated clay, tuffacious shale	Poorly fossiliferous, Scanty spores and pollen

12.5.3 Mainland Gujarat

In mainland Gujarat, the outcrops are few but in subsurface a near complete sequence of Paleogene and early Neogene rocks are encountered in the boreholes. The Paleocene is represented by Vagadkhol Formation which is an equivalent of Matanumadh Formation of Kutch. This is followed by Dinod Formation comprising limestone and bearing fossils indicative of Middle to Late Eocene age. Outcrops of Oligocene rocks are not exposed but they are present in subsurface where they have been designated as Tarkeshwar Formation. Dinod Formation is overlain by Babaguru and Kand Formations. The latter is highly fossiliferous and in addition to mollusks bears species of *Lepidocyclina, Myogypsina* and *Austrotrilina* indicative of

Burdigalian age. Kand Formation is overlain by Jhagadia Formation comprising sandstone and conglomerate. Jhagadia Formation does not bear any fossils and is presumed to be a fresh water deposition. In subsurface Jhagadia Formation is overlain by marine, fossiliferous Broach Formation comprising brown claystone and sandstone. The fossils comprising species of foraminifers *Rotalia, Elphidium, Bolivina* and *Cibicides* indicate a Pliocene age for this formation.

Borehole data from mainland and offshore regions available from drilling for hydrocarbons has revealed a thick sequence of marine Cenozoic rocks ranging from Paleocene to Pliocene.

12.6 PALEOGENE-EARLY NEOGENE SEDIMENTATION IN SOUTHERN BASINS

12.6.1 East Coast

In Tamil Nadu and Pondicherry (Puducheri) the Cretaceous marine sequence is unconformably overlain by Ninyur Formation (see Table 10.4). This 70-metre-thick formation comprises lower fossiliferous limestone, middle crystalline shelly limestone and upper argillaceous nodular limestone. The formation is highly fossiliferous especially the lower and middle part. The foaminifers include a wide variety of miliolids, notably *Triloculina* and *Quinqueloculina*, and planktonic foraminifer *Globorotalia.* It also bears a rich algal flora comprising *Dasiclodella savitrae, Acicularia dyumatsenae* and *Parachaetetes asvapathi* etc. The other fossils include the cephalopod *Nautilus danicus* and ostracodes *Bairdoppilata talukdarai* and *Xestoleberis rupnarayanalurensis.*The fauna and flora indicates a Danian age for the formation that could have been deposited in shallow marine or brackish water.

In Tamil Nadu, the continental Miocene is represented by rocks that may have originally been deposited as a continuous cover over all the older formations but have since been dissected by streams rendering them as isolated outcrops. This formation is known as Cuddalore Sandstone and it has a widespread distribution from Pondicherry (Pudducheri) to Thanjavur and Sivaganga. It comprises coarse to fine grained sandstone with bands of conglomerate and inter-calated clay. It indicates deposition in fluviatile, lacustrine and estuarine conditions. There is also a deltaic influence in the sediments. The formation bears the famous Neyveli lignite field from which lignite has been mined from several years. Silicified angiospermous and coniferous tree trunks abound in the formation. Palynological dating suggests a Miocene age for the formation.

The equivalent of Cuddalore Sandstone in the Godavari basin of Andhra Pradesh is Rajahmundry Formation comprising sandstone, conglomerate and claystone. This formation overlaps Deccan Trap and has a maximum thickness of 390 metres. The formation may have a long range and was probably a result of several phases of deposition and erosion. This also bears some lignite bands that contain palynomorphs and indicate the same age as that of Cuddalore Sandstone.

12.6.2 West Coast

Drilling in the offshore region of the coast of Kerala has shown the presence of marine Paleocene–Eocene sediments. There are no onshore outcrops of equivalent age. The oldest onshore outcrops of Cenozoic age constitute the Quilon Formation followed by Warkalli Formation having a thickness of 70 and 80 metres, respectively. Quilon Formation consists of fossiliferous limestone, sandstone and clay that bear mollusks, corals, foraminifers and ostracodes indicative of a Miocene age. Warkalli Formation consists of sandstone, sandy clay and a band of lignite with palynofloral data indicative of a Miocene age. The deposition is indicative of shoreline littoral conditions. The borehole data in the area has indicated that Quilon Formation and Warkalli Formation may be lateral equivalents and deposited at the same time in shallow marine conditions for the former and in extremely shallow coastal condition with paralic nature of sedimentation for the latter.

12.7 CENOZOIC ROCKS OF ANDAMAN AND NICOBAR ISLANDS

The group of Andaman-Nicobar Islands displays an almost continuous sequence of Cenozoic marine sedimentation, barring a few breaks. The Cretaceous rocks in the area constituting ophiolites, jasper and radiolarian chert are overlain unconformably by sediments referred to as Baratang Formation. This formation consists of grey shale, grey limestone and siltstone and bears Paleocene foraminifers. Baratang Formation is unconformably overlain by approximately 3000-metre-thick suite of rocks comprising conglomerate, grit and greywacke referred to as Port Blair Formation. This formation has a widespread distribution and bears an assemblage of radiolarians indicative of Middle Eocene age. Port Blair Formation is unconformably overlain by a suite of rocks constituting Archipelago Group. This group has been further divided into several formations. Archipelago Group consists of various kinds of rocks namely limestone, chalk, calcareous mudstone and calcarenite. Associated with these rocks are numerous volcanic ash layers and tuff. These are the traces of an intense episode of volcanic activity that commenced in south-east Asia during Early Miocene and continued up to Middle Miocene. Archipelago Group ranges in age from Late Oligocene to Pleistocene. Studies on foraminifers have indicated that there is a continuous marine sequence which can be dated on the basis of range zones of distinctive planktonic species that correspond with the datum levels as available from the Indian Ocean obtained through deep drilling. In this respect, Andaman-Nicobar is the only area in India where such a continuous Cenozoic marine sequence is available. Moreover, this is also the only region in India where marine Pleistocene rocks are exposed and where Pliocene–Pleistocene boundary can be fixed in a marine sequence (see Sec. 12.8.2).

12.8 PLIOCENE–PLEISTOCENE BOUNDARY

12.8.1 Global Status

For many years, there has been some debate about the placement of this boundary. The problem arose mainly because earlier the Cenozoic was divided into two Periods namely Tertiary and Quaternary, the former including Paleocene, Eocene, Oligocene, Miocene and Pliocene and the latter Pleistocene and Holocene. The boundary between Tertiary and Quaternary was accordingly the same as between Pliocene and Pleistocene. This boundary was marked by a distinct faunistic change especially among mammalian fauna on land. It was also marked by the advent of first phase of ice age. However, following the magnetic polarity studies that helped in precise dating and the study of the marine fauna in the Vrica type section (Calabria, southern Italy) and various submarine borehole studies, the boundary between Pliocene and Pleistocene was marked at the top of Olduvai subchron or beginning of C1n polarity chron dating 1.81 Ma. This is not the same as the original boundary between Tertiary and Quaternary which is dated around 2.6 Ma. Quaternary, therefore, was not considered to commence from the beginning of Pleistocene but earlier. For many years, the Pliocene-Pleistocene boundary was not considered to be the same as the Neogene-Quaternary boundary. However, in recent years this opinion has been reversed by the International Commission of Stratigraphy and Quaternary has been given the status of a System/Period succeeding the Neogene and the Neogene-Quaternary boundary has been considered as the same as the Pliocene-Pliestocene Boundary dated around 2.58 Ma.

12.8.2 Pliocene-Pleistocene boundary in Indian stratigraphy

Pliocene-Pleistocene boundary in India has not been clearly demarcated in line with the accepted Global Stratotype Section and Point (GSSP) at Vrica (Italy). The only section where marine sequence at this level is available is Andaman-Nicobar Islands (see Sec. 12.7). While this section has been investigated biostratigraphically and range zones have been identified on the principle of first appearance datum and last appearance datum, there is no clear zonal identity either at 1.81 Ma or 2.58 Ma level, but at the level 1.9 Ma that marks the FAD of *Globorotalia truncatulinoides.* This is slightly lower than the earlier accepted boundary concept and much higher from the presently accepted position of the boundary. In the fresh water sequence of Siwalik Group the boundary can be fixed on the basis of first appearance of E*quus* and *Cervus* that nearly coincides with the base of Pinjor Formation (see Table 13.1). This dates broadly around 2.48 Ma as per magnetic polarity chron. However, the Pliocene-Pleistocene boundary would be at slightly lower position, dating 2.58. Faunistically there is no significant variation at this level.

12.9 PLEISTOCENE–HOLOCENE DEPOSITS

12.9.1 Distribution

Except for Andaman-Nicobar Islands (see Sec. 12.8.2) and to a limited extent the coastal area of Saurashtra (Gujarat), there are no marine Pleistocene deposits in India. Fluvial, glacial, lacustrine and aeolian deposits during Pleistocene and Holocene are widely distributed throughout the sub-continent. In the Himalayan region, the Siwalik Group extends into Pleistocene (see Chapter 13). In addition, Pleistocene and Holocene sediments are widely distributed in intra-montane basins of Kashmir and Nepal. The river and stream valleys of higher Himalaya display the evidences of glacial events in the terraces of Pleistocene and Holocene age that have subsequently been dissected by stream waters following the final phase of Himalayan uplift. The final phase of Himalayan uplift took place in post-Siwalik times due to which Siwaliks got uplifted and folded, creating a new basin of deposition in front of the Himalayan chain. The Indo-Gangetic alluvium got deposited in this basin with sediments derived from Himalaya through a number of subsidiaries of Indus and Ganga river systems.

Aeolian deposits of Pleistocene-Holocene times cover a wide area constituting dunes and *kankar* in what is known as Thar desert that covers most of western Rajasthan and Sindh area of Pakistan. In Peninsular India, the deposits are widespread around river valleys notably those of Narmada, Tapti, Godavari, Krishna and Penner. Most of the older rock formations of peninsular India are covered by laterite that has been the product of intense subaerial weathering during Pleistocene-Holocene times.

12.9.2 Karewas of Kashmir

Following several phases of Himalayan orogeny an intra-montane basin was formed between the Pir Panjal Range to the south and Great Himalayan Range and North Kashmir Range to the north, in what is now the Kashmir valley. This basin became the site of fluvial, lacustrine, glacial and aeolian deposits during Late Pliocene to Early Holocene. The resultant sediments constitute what is referred to as Karewa Group (based on the local name of the flat hillocks constituting this group). The bulk of Karewa sediments, especially the lower part, are of lacustrine origin as a single lake covered the entire valley and was fed by the streams from surrounding mountains. The length of the valley is approximately 140 kilometres and has an average width approximately 30 kilometres. Karewa Group of rocks covers almost the entire valley resting on rocks of ages varying from Cambrian to Triassic and are horizontally displaced for the most part or mildly dipping along the margins of the valley. Along the main river Jhelum and its tributaries, the Karewa sediments are dissected and Holocene alluvium is deposited in their flood plains that rests on the Karewas sometimes with a perceptible unconformity. The thickness of Karewa Group is approximately 1300 metres.

Karewa Group is divisible into three formations namely Hirpur, Nagum and Dilpur. Hirpur Formation is well developed towards the south-western part of the valley where it has suffered uplift in the final phase of Himalayan orogeny during Pleistocene. Thus, it also occurs at higher altitudes especially along Pir Panjal range and shows an angular unconformity at many places and a disconformity elsewhere with overlying Nagum Formation. Hirpur Formation was mainly deposited in the lake and lithology comprises light grey sandy clay, green to purple sand with occasional bands of conglomerate. Towards the marginal side around Nichihama and Raithan there are seams of lignite that must have developed in swamps that bordered the lake. Nagum Formation covers most of the central part of the valley and comprises grey and ochre coloured sandy clay, ochre and cream coloured marl. Dilpur Formation comprises loamy and loess deposits. This formation was deposited after the obliteration of the lake system and its drainage. Thus, some part of this formation comprising loess is of aeolian origin.

There are evidences of glaciation in Karewa sediments in the shape of conglomerate and varved clays. However, the exact correlation of the same with different phases of Pleistocene glaciation is still a matter of doubt. The gravel bed below the loess of Dilpur Formation (Malshaibagh conglomerate) represents the commencement of the glacial events. Four to five interglacial palaeosols can be seen in the loess deposits

The bulk of Hirpur Formation is considered to be of Pliocene age extending into Early Pleistocene. On the basis of thermoluminiscence dating for the loess deposit, the base of Dilpur Formation is presumed to be dating around 300 Ka or the upper part of Middle Pleistocene. The Karewa deposition probably stopped at the end of Pleistocene. The alluvial tracts and flood plains of Jhelum and its tributaries constitute the Holocene deposits.

Karewa Group bears fossils of plants, fishes and mammals. Plant fossils in the form of leaves and fruits are extremely common in Hirpur Formation, especially in areas where lignite is present. They include most of the forms that are found nowadays in the valley although the deposits are sometimes at higher elevations where they do not exist now. This is evidently a result of the uplift that Hirpur formation has suffered. The forms include *Cedrus, Pinus, Salix, Juglans, Betula, Trapa* among others. Remains of modern fish *Schizothorax* have also been reported. Remains of micro-mammals include rodents and insectivores while larger mammals include all present-day forms such as *Bos, Sus, Cervus* and *Felis*. Most interesting are the fossils of *Elephas* which is no longer found in Kashmir at present.

12.9.3 Indo-Gangetic Plain

After the uplift and folding of the Siwalik Group during the latest phase of Himalayan orogeny a foredeep was formed in front of the Himalayan mountain chain that became the site of deposition of sediments derived from the mountain through the agency of rivers belonging to the Indus and Ganga river systems. The basement of this foredeep was in all probability the frontal margin of the Indian plate bearing Proterozoic rocks. As the plate went on subducting (see Chapter 11) the foredeep continued to sink throughout Late Pleistocene and Holocene. Consequently, an enormous thickness of sediments was deposited extending from Punjab in the west to Bengal in the east. This constitutes what is known as Indo-Gangetic plain that separates the Himalayan range from Peninsular India. Age wise it ranges from Late Pleistocene to Holocene.

The sediments of Indo-Gangetic plain are mainly alluvial with occasional lacustrine and aeolian elements. Bordering the Siwalik rocks to the north there are older fans and piedmont especially towards the western side in Punjab and Haryana. This is overlain by alluvium and sometimes capped by aeolian sediments. There is apparently a faulted contact between Siwalik rocks and these older units and it is presumed that neo-tectonic movements have resulted in Siwaliks overriding the Indo-Gangetic sediments along this fault. In the central part, the older alluvium consists of gravels that are followed by alluvial fan deposits and river terraces. Towards the eastern side the Ganga and Brahmaputra constitute a widespread basin where thick alluvial deposits occur in Bengal and Assam. While towards north the flood plain deposits are overlapped by piedmont alluvium, to the south the Indo-Gangetic plain merges with mangrove swamps and mudflats that occasionally bear estuarine foraminifera.

12.9.4 Thar Desert

South-west of Indo-Gangetic plain, the Thar Desert occupies a substantial area on the north-western margin of peninsular India. The onset of Pleistocene marked a major climatic change from humid–sub-humid to arid condition. Such dryland environment prevailed for a considerable amount of time with occasional onset of relatively wetter periods. This was one of the reasons for the increase in intensity of the desertification of this region. The desert sand occupies a substantial area of western Rajasthan where large sand dunes of aeolian origin constitute the main landscape in the region. The history of Thar Desert is at least 150 ka old. Periods of enhanced sand mobilisation and dune sand accumulation during the Late Pleistocene were luminescence dated to be ca 100, 75, 55, 25–30 and 10–14 ka. There are at least three generations of near surface dune deposits. The oldest generation constitutes the consolidated fine sand deposits that are followed by semi-stable pedogenised dunes and the youngest active aeolian dunes. On the basis of archaeological evidence, the oldest exposed dune unit could be at least of about ca 14 Ka before present (B.P.). The dune sand is mostly derived from alluvium though there has been some contribution from the Arabian sea. Below the sand dunes as per the exposures along the Luni river basin and borehole data, well sorted aeolian calcareous sand constitutes the Pleistocene deposits that are indicative of semiarid-arid conditions during that time span. These deposits also bear Middle Palaeolithic stone implements. The arid conditions are gradually replaced by sedimentation of a more moist type eastwards in Rajasthan-notably, across the Aravalli range.

12.9.5 Saurashtra (Gujarat)

Coastal region of Saurashtra is one of the areas where sediments of Pleistocene-Holocene age are deposited under the combined effect of marine, fresh water and aeolian conditions. Early Pleistocene comprises gravel and sand which is followed by the Miliolite Group of the Middle-Late Pleistocene age and the Chaya Formation of Holocene age. Of special interest is the Miliolite Group (also referred to as Porbandar Stone) that is primarily a shell limestone made up of cemented tests of foraminifers of the Miliolidae family, together with oolites and calcareous sand grains. Also present in the formation are numerous tests of other benthic foraminifera notably those belonging to the families Rotalidae, Elphididae and Nubeculariidae. Broken fragments of mollusks, bryozoans, corals, echinoids and ostracodes are also present. Most of the tests are broken, rolled up and abraded making the identification difficult. While the material is undoubtedly of marine origin, bulk of it appears to have been deposited under aeolian conditions especially in the inland areas. However, nearer to the coast, the basal parts of the formation do not display any evidences of transportation and could have been deposited in shallow marine conditions.

12.9.6 Narmada Valley

The Narmada valley has been a site of rifting in cental India (see Chapter 14) and is bound by faults to the north and south. These faults got re-activated during Pleistocene resulting into the formation of a basin where Pleistocene-Holocene sediments of the order of approximately 400 m got deposited. This is one of the most important central Indian river basins where such a thickness of these sediments is exposed. The Pleistocene sediments have been divided into five Formations namely, the Pilikarar, Dhansi, Surajkund, Baneta and Hirdepur. The Holocene comprises river terraces at various levels. Of special interest is the Baneta Formation that bears volcanic ash, which has been correlated to the younger Toba eruption that is dated to approximately 70 Ka B.P. or beginning of the Late Pleistocene.

The Pleistocene deposits of the Narmada valley are repositary of a wide variety of mammalian fossil bones, a hominid skull and the Stone Age implements. The fossils are evidently reworked as they are found in channel deposits. The mammalian fossils include *Sus namadicus, Elephas hysudricus, Stegodoninsignis-ganesa, Hexaprotodon palaeindicus, Equus namadicus, Bos* and *Cervus*. The human skull belongs to *Homo erectus* and on the basis of fluorine dating indicates late Middle Pleistocene age. The Palaeolithic stone implements include choppers, hand axes, flake-blades, cleavers and scrapers.

■■■■

13 The Siwalik and Related Sedimentation

13.1 NOMENCLATURE AND DISTRIBUTION

The emergence of Himalayan range following the Early Miocene phase of uplift resulted into the formation of a foreland basin in front of the newly emerged mountain chain. This basin became the site of extensive deposition of fluviatile sediments derived from the uplifted regions of the mountain chain. The basin extended throughout the length of the Himalayan mountain chain from Sindh in west to Arunachal in the east. The suite of sedimentary rocks deposited in this basin has been named as the Siwalik Group. The name is derived from the range of hills stretching from Jammu to foothills of Nepal that comprises series of parallel ridges and is principally made up of these rocks.

The equivalents of these rocks in Sindh and Baluchistan are referred to as Manchhar Group that is partly deposited in fluviatile conditions and partly in marine and estuarine conditions. Except for Sindh, the Siwalik rocks throughout are deposited in fresh water conditions. While almost all units of the Group outcrop along the Himalayan foothills, the best exposures of the lower and middle part are known from the area between Salt Range and Rawalpindi, including Khaur and Dhulian domes, and Potwar plateau of Pakistan from where the nomenclature of most of the formations of the Group is derived.

While the main Siwalik outcrop extends in the foothills of Arunachal in the east, the equivalents of the same in Assam are referred to as Tipam Formation (see Sec. 12.4.1). However, this formation does not have any physical continuity with main Siwalik rocks, since they are separated by a thick cover of alluvium. It is possible that Tipam Formation may be a continuation of Siwalik rocks under the alluvial cover since they correspond in lithology and age. Likewise, the Irrawady Group of Mayanmar (Burma) may also represent the extension of Siwalik Group.

The thickness of the Siwalik Group is variable from place to place. However, the greatest thickness is towards the western part of Himalayan foothills and in Pakistan. The total calculated thickness is over 6,000 metres, though in no single section is this thickness exposed.

13.2 CLASSIFICATION

The classification of Siwalik Group of rocks was first proposed by Pilgrim in 1910 and this classification has not been changed since except for some minor modifications, notably the substitution of Kamlial in place of Lower Manchhar. However, the suffix of 'stage' for each of the units as proposed by him is presently being replaced by Formation, in keeping with the code of stratigraphic nomenclature. Because of the monotonously uniform nature of lithology of Siwalik Group and local variations, the classification is not so much based on lithological parameters as on the fossil content. Accordingly, it is not strictly in line with the lithostratigraphic principles. The classification is, therefore, of limited value in unfossiliferous Siwalik sections and the correlation in these cases becomes arbitrary. The classification can be broadly summarised as shown in Table 13.1.

Table 13.1 Classification of Siwalik Group

SUB-GROUP	FORMATION	DISTINCTIVE LITHOLOGY	DIAGNOSTIC FOSSILS
UPPER	Boulder Conglomerate	Boulders, cobbles and pebbles, predominantly of quartzite with occasional granite, trap and limestone in a matrix of coarse sand.	*Elephas hysudricus, Cervus, Camelus.*
	Pinjor	Massive, greyish white sandstone, interbedded with reddish-brown clay.	*Elephas, Equus sivalensis, Archidiskodon planifrons, Leptobos, Cervus.*
	Tatrot	Friable grey sandstone with thin clay bands	*Stegodon, Hipparion, Hexaprotodon sivalensis.*
MIDDLE	Dhok Pathan	Feldspathic sandstone, interbedded with conglomerate and mudstone.	*Ramapithecus, Gigantopithecus, Hipparion, Giraffokeryx, Sus.*
	Nagri	Thick cross-bedded sandstone with interbedded siltstone.	*Sivapithecus, Hipparion.*
LOWER	Chinji	Red mudstone and siltstone alternating with bands of coarse sandstone and conglomerate.	*Deinotherium, Sivapithecus, Trilophodon, Progiraffa.*
	Kamlial	Sandstone alternating red mudstone	*Deinotherium, Anthracotherium.*

13.3 TIME OF SIWALIK DEPOSITION AND CORRELATION

Since all the type sections, with the lone exception of Pinjor, are in Pakistan, it is difficult to correlate the Indian sections especially when the fossils are erratically preserved or totally absent in a sequence. The lithology does not offer much help since there is considerable lateral variation. In recent years, the magnetic stratigraphy has provided a useful mechanism for correlation for the total length of Siwalik sequence. Magnetic polarity studies have been undertaken extensively in Potwar region and other areas of Pakistan. Corresponding studies have also been undertaken in Indian sequences, notably in Hari Talyangar area in Himachal Pradesh, Pinjaur near Chandigarh and Jammu Hills.

These studies are, however, restricted to Middle and Upper Siwalik. Based on these studies and polarity chrons established thereby, the age of the different units can be estimated with some degree of precision. The results are broadly summarised in Table 13.2.

Table 13.2 Age Relationship of Various Formations of Siwalik Group

Division	Formation	Epoch	Polarity chron	Zone (Number)	Age (Ma)
UPPER	Boulder Conglomerate	Pleistocene			
	Pinjor		Matuyama		2.47
	Tatrot	Pliocene	Gauss	C2A (=3)	5.1
MIDDLE	Dhok Pathan		Gilbert	C3 (=4)	8.5
				C3B(=6)	? 10.8
	Nagri			C4A (=8)	
LOWER	Chinji	Miocene		C5A(=10)	14.3
				C5B (=15)	? 18.3
	Kamlial			C5D (=17)	

13.4 LITHOLOGY AND PETROGRAPHY

Throughout the thickness of Siwalik rocks the sandstone-mudstone alternation is a common phenomenon. However, pebbly sandstone and conglomerate predominates towards the upper part of the sequence. The sequence shows enormous local variation in the thickness of individual alternating beds. Most of the units are time transgressive and hence generalisation on a regional basis is not possible.

In the Lower Siwalik the percentage of sandstone varies from 73 to 85% in Kamlial Formation and its equivalents, the remaining part being the mudstone. In Chinji the mudstone and claystone is the more dominant unit. In Mianwali region of Pakistan the sandstone beds in Kamlial Formation form long strike ridges that are frequently interbedded with pseudo-conglomerates. The Chinji Formation here comprises the highest percentage of mudstone (approximately 73%). The Lower Siwalik in Jammu Hills display cyclic sedimentation comprising sandstone followed successively by siltstone, massive claystone, concretionary claystone, laminated claystone and olive green claystone. The sequence is characterised by bands of maroon to reddish-brown mudstone bearing calcrete nodules. A more or less similar sequence can be seen in Kumaun Hills and Nepal. However, in the Indian part of Siwalik it is difficult to distinguish between the Kamlial and Chinji units. The Lower Siwalik is characterised by the heavy mineral assemblage comprising opaque minerals, garnet, tourmaline, epidote, staurolite, zircon and rutile.

Middle Siwalik consists of medium-grained, thick cross-bedded sandstone alternating with mudstone. The lower unit comprising Nagri Formation is characterised by dominance of multistoried sand bodies. Concretionary layers are abundant in the sequence. The upper part of Middle Siwalik corresponding to Dhok Pathan Formation is composed of gray, micaceous, feldspathic sandstone, pebbly sandstone and occasional conglomerate beds. In the Indian part the equivalents of Nagri comprise thick sandstone with interbedded siltstone and shale, whereas the equivalents of Dhok Pathan comprise sandstone, pebbly sandstone, occasional conglomerate and varicoloured mudstone. In Jammu Hills the lower part of Middle Siwalik sandstone is profusely cross-bedded and identical to that of Potwar region. The heavy mineral assemblage shows an influx of kyanite, in addition to others present in Lower Siwalik rocks.

The Upper Siwalik is best exposed in Jammu Hills and Chandigarh region, the latter being the type area for Pinjor Formation. In Jammu Hills, the equivalents of Tatrot and Pinjor have been named as Parmandal Sandstone and Nagrota Formation, respectively. The boundary between the two may not exactly coincide with the Tatrot–Pinjor boundary but shows a distinct lithological division that is mappable. Parmandal Sandstone comprises coarse-grained thick cross-bedded sandstone alternating with reddish brown clay. Nagrota Formation shows domination of siltstone and clay over sandstone. It also bears a band of bentonised tuff, indicative of a volcanic activity in some neighbouring region. Bands of volcanic tuff are also known in Jhelum re-entrant area in Pakistan in Upper Siwalik sequence. In Chandigarh region, the lower part corresponding to Tatrot comprises alternation of medium to coarse grained sandstone with variegated mudstone while in Pinjor single storied sandstone is the dominating lithological element. Boulder Conglomerate unit is universally present throughout the sequence though it is variable in thickness from place to place. The Upper Siwalik rocks bear a complex suite of heavy minerals with the new elements like hornblende, sillimanite and andalusite added to the other minerals occurring in older rocks.

The mineralogy of Siwalik sandstones is unusual. On an average it comprises 40% quartz, 15% feldspar, 6% mica and from 35 to 40 % metamorphic rock fragments. Some sandstones are dark grey in colour and on the basis of composition can also be classed as graywackes.

13.5 FAUNA

Since Siwalik rocks represent a time period when mammalian evolution was at its acme, profuse mammalian fossils are known from these rocks in addition to reptiles, mollusks, plant fossils, spores and pollen grains. The evolutionary trends, faunal migration and dispersal of mammals during the time span of Siwalik sedimentation also throws light on the palaeogeography. The mammals comprise elements of

Carnivora, Proboscidae, Suidae, Equidae, Giraffidae, Bovidae, Rhinoceratidae, Anthracotheriidae, Rodentia and Primates. However, the fauna is restricted to some pockets and is not universally distributed. This is understandable because bulk of Siwalik rocks were deposited in stream and river channels, in addition to flood plains. As a result, the bones of the animals got washed off in these channels and were rapidly covered up by sediments. Accordingly, the fossils are found only in these channels while bulk of the sediment is generally unfossiliferous.

While Siwalik rocks bear diverse species, most of the elements are not of much biostratigraphic or phylogenetic significance. However, some forms are useful as horizon markers (see Table 13.1) while others are significant in determining the migration pattern, palaeogeography and palaeoclimate. Globally Early and part of Middle Miocene has been a period of arid climate and Indian sub-continent was no exception. Because of limited vegetation the expansion of mammals was also hampered. This pattern changed towards the later part of Miocene and there was a radical faunal change and advanced forms appeared rapidly in Chinji, Nagri and Dhok Pathan Formations. This was because of restoration of humid conditions which resulted in expansion of grasslands. An explosive evolution and diversification of bovids, cervids, giraffids, proboscids and other herbivorous elements were followed by expansion of carnivores.

Of special interest from a point of view of phylogeny and migration are the members of equidae family. *Hipparion*, the three-toed horse, seems to persist throughout Middle Siwalik extending into early part of Upper Siwalik. In the remaining part of Upper Siwalik *Equus* replaces it. Since *Hipparion* is not a direct ancestor of modern single-toed horse *Equus,* it is believed that the latter migrated to India presumably from North America, while *Hipparion* became extinct towards the beginning of Pleistocene epoch. Proboscidians constitute another group of significance to Siwalik stratigraphy. *Deinotherium* is an important element of Lower Siwalik rocks together with various types of mastodons. Middle Siwalik rocks do not contain any significant proboscidians and the intermediate stages of proboscidian phylogeny are not known here. *Stegodon* appears at the beginning of Upper Siwalik in Tatrot rocks and is succeeded in Pinjor by *Elephas.* The simultaneous occurrence of *Equus, Elephas, Bos* and cervids with antlers at the beginning of Pinjor is a horizon marker that has been dated at 2.47 Ma marking the beginning of Matuyama polarity zone.

Another interesting group of fossils belong to primates. *Sivapithecus* appeared in Lower Siwalik and disappeared at the beginning of Middle Siwalik. Of special significance are the genera *Ramapithecus* and *Gigantopithecus* that appeared in Middle Siwalik in rocks equivalent to Dhok Pathan Formation in Hari Talyangar area of Himachal Pradesh. The former is considered to be a direct ancestor of the modern human family. The Indian species *Ramapithecus punjabicus* and *Gigantopithecus bilaspurensis* have been dated at 6.3 Ma corresponding to C3B (=6) polarity chron.

The Siwalik rocks also bear a wide variety of bovids, cervids, hippopotamids and giraffids in addition to carnivores. The genera like *Giraffokeryx, Anthracodon, Hexaprotodon, Sivatherium, Crocuta* are of special significance as horizon markers in different sections.

13.6 SIWALIK SEDIMENTATION AND TECTONICS

As has been mentioned earlier, the bulk of Siwalik sedimentation is of fluvial type. The stream channels have primarily flown from north to south, but variations in the direction and nature of channels and the orientation of associated flood plains are a common feature. At places the sedimentation and deposition has taken place as braided streams and alluvial fans. The cyclicity in sedimentation indicates a balance between the basinal subsidence and fluvial deposition.

The Cenozoic stratigraphy of Himalaya is directly linked with northward migration of Indian plate and its eventual collision with Asiatic plate (see Chapter 14). The evolution of the Siwalik basin represents a phase in this migration when the remnants of the Tethys sea had finally receded and the collision resulted in emergence of Himalayan orogen as an area of positive relief. The newly emerged orogen became the provenance for the sediments that were deposited in the Siwalik basin. The Siwalik deposition has been considered to be representing the molasse facies. The large thickness of Siwalik sediments cannot be explained without taking recourse to basinal subsidence that must have been a result of post-collision tectonics. The tectonic pattern has been the one where uplift due to the Indian plate going under the Asiatic plate was accompanied by formation of a foredeep in front of the mountain chain that became the site of deposition of sediments derived from this chain. As the uplift continued the older Siwalik sediments also got folded and uplifted resulting in the migration of the foredeep further southwards where younger Siwalik rocks got deposited. The entire thickness of Siwalik rocks was eventually involved in the Himalayan orogeny and got uplifted and folded becoming an area of positive relief. As a result, in post-Siwalik times the foredeep moved further southwards and the deposition of Indo-Gangetic alluvium is the later example of the foredeep deposition similar to that of Siwalik rocks, a phenomenon that is continuing to this day.

During the uplift, Siwalik rocks were involved in folding and were overthrust by older rocks in the central and eastern part of India, along what is known as the Main Boundary Fault[1]. This fault splits into two or more faults to the west whereby the older Cenozoic rocks (Murree, Dagshai–Kasauli, and Sabathu and their equivalents) overthrust the Siwalik along one of the faults and are in turn thrust over by the pre-Cenozoic rocks along another thrust. The older Siwalik rocks are involved in tight folds that are sometimes inverted especially towards the eastern part. However, to the west there are open folds that are dissected by high angled reverse faults. There is reason to believe that Siwalik rocks themselves are overthrusting the younger Indo-Gangetic sediments along certain sectors. This Himalayan frontal thrust is currently in the process of evolution and neotectonic activity in the region is associated with the generation of this thrust and the associated cross faults.

■■■■

[1]Main Boundary Fault is different from the Main Boundary Thrust. The latter separates the Pre-Cenozoic rocks from Cenozoic rocks whereas the former separates older Cenozoic rocks (Murrees and equivalents) from Siwalik Group

14 Tectonic Evolution of Indian Sub-Continent

14.1 EVOLUTION OF THE INDIAN SHIELD

The Earth's crust has passed through various cycles of orogeny, metamorphism, granitisation and magmatic activity. There has also been a continuous movement of crustal plates, their split through rifting and juxtaposition through collision. This phenomenon has been going on for billions of years since the formation of the crust through Archean and Proterozoic resulting into the consolidation of the shield areas of the world.

The Indian shield was also formed by a prolonged period of several cycles of orogeny through Archean and Proterozoic times. These orogenies and related metamorphism and igneous activity resulted into nucleation of the different cratons and their final consolidation. In a generalised way, the main nuclei around which the consolidation took place were the different cratonic complexes namely Dharwar, Bastar, Singhbhum, Bundhelkhand and Aravalli. These cratons passed through several orogenic cycles resulting into formation of varied lithological complexes that indicate varying ages of crystallisation.

While the number and precise timing of the orogenic cycles cannot be fixed in all cases, the various Precambrian cycles can be broadly grouped into four sets. The first and the oldest set comprise several cycles with a long range of time dating from 3600 to 2900 Ma. The next set covers a relatively shorter span ranging from 2800 to 2500 Ma. These are the two major sets of the Archean era. The other two sets are of Proterozoic span and show age ranges of 2400–1800 Ma and 1400–750 Ma. The Proterozoic sets also show signatures in the Himalayan region. In addition, there is evidence of 600–500 Ma orogeny, especially in Himalaya, that is usually considered as related to Pan-African orogeny which affected the entire Gondwana supercontinent. Table 14.1 summarises the different sets and the stratigraphic units related to them.

Table 14.1 Sets of Precambrian orogenic cycles in Indian Subcontinent

Time span (Ma)	Dharwar	Bastar	Singhbhum	Bundhel-khand	Aravalli	Himalaya
1400-1900			Younger Granitoids and Newer Dolerite		Delhi Supergroup	
2400-1800		Sakoli Group	Dalma Group	Mafic Dykes	Aravalli Supergroup	Vaikrita Crystallines
2800-2500	Dharwar Supergroup Closepet Granite	Bengpal Group	Dhanjori Group	Younger Granites	Banded Gneissic Complex	
3600-2900	Peninsular Gneiss Gorur Gneiss	Older Granite Complex	Iron Ore Group Older Metamorphic Group	Older Granite	Banded Gneissic Complex	

14.2 INDIAN SHIELD BEFORE AND DURING PALEOZOIC

14.2.1 Proto-Gondwana

Before the commencement of Phanerozoic, Indian subcontinent was a part of a super-continent designated as Proto-Gondwana. Proto-Gondwana itself was formed after the break-up of a Proterozoic super-continent named Rodinia. Rodinia was formed approximately 1.1 billion years ago, and it split into three super-continents 750–700 million years ago (Fig. 14.1) (Rodinia itselt was derived from several earlier supercontinents). The three supercontinents namely Proto-Laurasia, Proto-Gondwana and Congo craton moved away from one another. Proto-Lauratia further split up through rifting into Laurentia, Siberia and Baltica. By the beginning of Cambrian, the Proto-Gondwana comprising Australia, Antarctica, India, Madagascar, parts of Africa, South America, China, South Europe, Iran and Turkey was located on either side of equator and was separated from Laurentia by Proto-Tethys.

14.2.2 Proto-Gondwana through Paleozoic Times

The process of rifting within Proto-Gondwana continued throughout Paleozoic due to which parts of north and south China rifted away and headed northwards. This rifting resulted into the formation of a new ocean, the Palaeo-Tethys. The parts of south Europe also fragmented from Proto-Gondwana by the end of Silurian and moved rapidly northwards colliding with Baltica in Devonian. Proto-Gondwana suffered several cycles of orogeny apart from the rifting mentioned above. The orogeny that left its signature in the Indian subcontinent is the earlier phase of Pan African orogeny. The evidences of this orogeny are in the form of granite emplacement in parts of Himalaya dating approximately 500 Ma and the break in sedimentation and an angular unconformity from Late Cambrian to Middle Ordovician (see Sec. 7.5).

What was left of Proto-Gondwana after the breaking away of China and South Europe eventually came to be named as Gondwana supercontinent (earlier referred to as Gondwanaland). It moved rapidly southwards eventually touching the South Pole by Late Carboniferous.

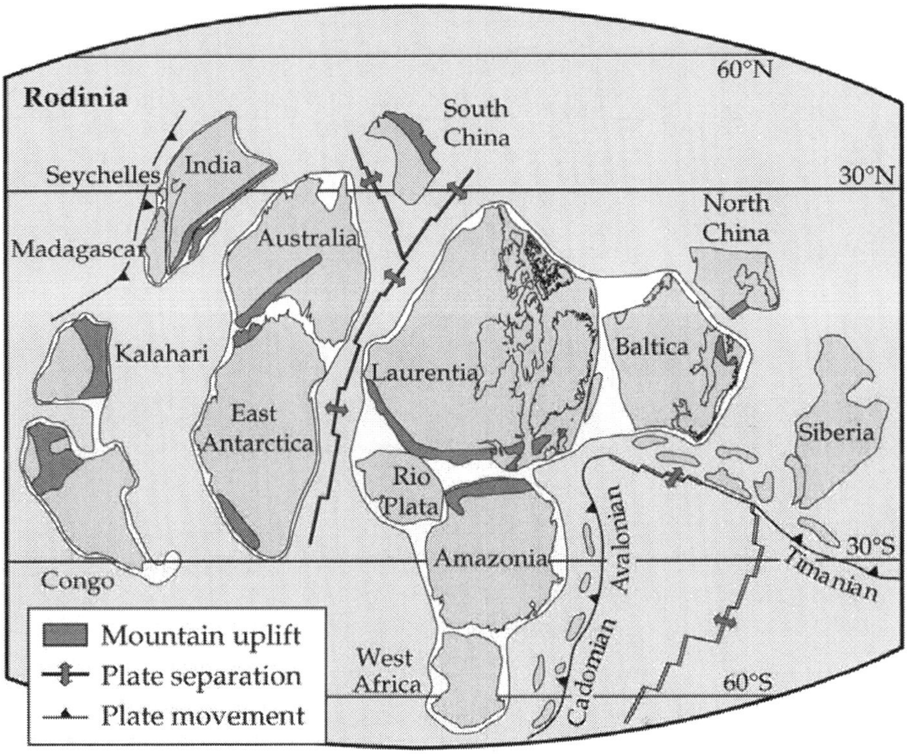

Fig. 14.1 Reconstruction of Rodinia Super-continent during its split approximately 750 Ma.
(Source: *The Encyclopedia of Earth, Ancient Earth* Edited by Arnold J. Bloom)

14.3 GONDWANA SUPERCONTINENT

14.3.1 Configuration

As mentioned above, the name Gondwana supercontinent was given to the descendent of Proto-Gondwana after China and south Europe separated from it. This supercontinent comprised Australia, New Guinea, Antarctica, India, Madagascar, Africa, South America, Arabia, Iran, Tibet and Turkey (see Fig. 14.2). It moved rapidly southwards and its eastern portion comprising India, Antarctica and Australia were headed towards South Pole by Carboniferous. During Late Carboniferous, while considerable portion of the supercontinent was located around South Pole, some parts extended further to the north almost nearing equator. The location of the part of it at the South Pole or extremely close to it resulted into a major ice age that engulfed this supercontinent by Late Carboniferous and Early Permian. The evidence of this ice age is widespread in India, Australia, Antarctica and South Africa.

The northern margin of Gondwana supercontinent was bounded by Neo-Tethys to distinguish it from Paleo-Tethys mentioned above. The total extension of Neo-Tethys is debatable. According to one opinion, it constituted an ocean of vast dimensions while for others it was only a thin linear epicontinental sea separating Gondwana from Laurentia on the one side and southern China on the other. However, recent data has shown that the former view is more plausible. During Late Carboniferous and Early Permian while most of the Gondwana was engulfed in an ice age, tectonic activity and island arc volcanism was an important event along the northern boundary of the supercontinent in Indian part (Panjal volcanics, Abor Volcanics). Within the Indian shield itself, the Proterozoic mobile belt regions notably Son–Narmada, Pranhita–Godavari got re-activated as rifts producing fault separated basins that eventually became sites of Gondwana sedimentation.

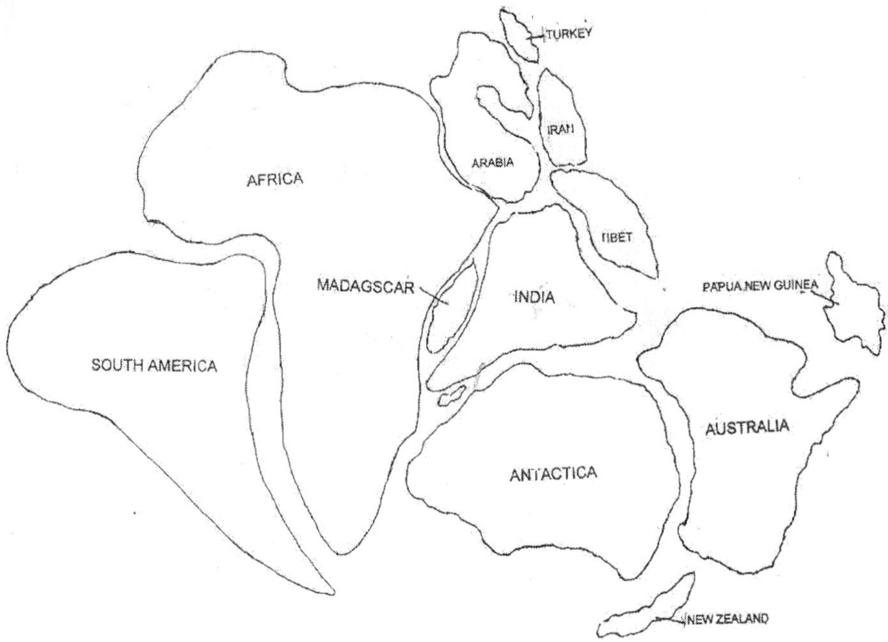

Fig. 14.2 Configuration of Gondwana Supercontinent

The various continents and landmasses constituting the Gondwana supercontinent bear an identical flora and fauna. In fact, it was the similarity of this flora and fauna in widely separated localities like South America, Africa, India and Australia that drew the attention of earlier geologists like Taylor and Wegener who proprounded the theory of continental drift to explain the phenomenon. By now, with the advent of the concept of plate tectonics, we have a better idea of plate movement and their break up and collision mechanism to explain the evolution of the Earth's crust in space and time.

14.3.2 Break-up of the Gondwana Supercontinent

The earliest split in Gondwana supercontinent started by breaking away of what is known as Cimmerian Plate in Early Permian, and its movement towards Eurasia and its collision. Part of the same split was that of the sub-plate constituting the Lhasa block (Tibet) that also drifted away and moved across the Paleo-Tethys for its rendezvous with south China which had earlier separated from Proto-Gondwana. These movements resulted in the closure of Paleo-Tethys and expansion of Neotethys.

The main break-up of Gondwana supercontinent commenced in Middle Jurassic (approximately 167 Ma ago) when eastern part of Gondwana comprising Antarctica, Madagascar, India and Australia separated from Africa and South America. Eventually South America drifted rapidly away from Africa opening up the South Atlantic Ocean. Madagascar and India separated from Antarctica and moved northwards opening up the Indian Ocean in the process. Subsequently the remaining eastern part of Gondwana also got dismembered in stages. Madagascar block got detached first followed by Seychelles Islands from India, the process starting around 120 Ma ago in Early Cretaceous and culminating by the end of Cretaceous. Australia also separated from Antarctica in Late Cretaceous approximately 80 Ma ago.

14.3.3 Flight of the Indian Plate to the North

After Madagascar and India separated, India continued to move northwards at a rate approximately 15 cm per year, which is perhaps the fastest rate at which any plate has moved, while Madagascar stopped and got locked to the African plate. During the flight of the Indian plate northwards it passed over various hot spots or plumes that can be seen today in the Indian Ocean. Many intraplate volcanic episodes in India are attributed to the activity due to these hot spots when Indian plate was passing over them. For example, the Rajmahal Trap, dated around 117 Ma is attributed to Kerguelen hot spot that is presently located near 50° S latitude. Likewise, Deccan Trap activity is attributed to Reunion hot spot located around 20° S latitude at present. There is a possibility that the location of hot spots themselves has changed during the passage of time. Nevertheless, the role of these hot spots in generating intraplate volcanic activity is widely accepted.

14.4 COLLISION AND SUBDUCTION

14.4.1 Closure of Neo-Tethys

The rapid movement of Indian plate northwards resulted into narrowing of Neotethys (see Fig. 14.3). This process continued through Jurassic and Cretaceous and by the time Seychelles fragmented from India and Deccan volcanic activity commenced along the western side of the plate, the Neo-tethys had narrowed down considerably. Along the northern boundary of the Indian plate deep trenchs resulted in the sedimentation of flyschoidal rocks in Late Cretaceous in association with deep marine oozes. On the northern side the Neo-Tethys was accreted to the margin of Eurasia where Dras–Kohistan volcanic arc was generated in an intraoceanic setting.

It is now widely accepted that the collision between the Indian Plate with southern Eurasia constituting the Karakoram–Lhasa Blocks resulting in the final closure of Neo-tethys occurred during Middle to Late Eocene dating around 50–45 Ma. This timing is deduced from several lines of evidence: notably, the ending of marine sedimentation in the Indus Suture Zone and the beginning of molasse facies and the initiation of major collision related thrust structures in Himalayan region.

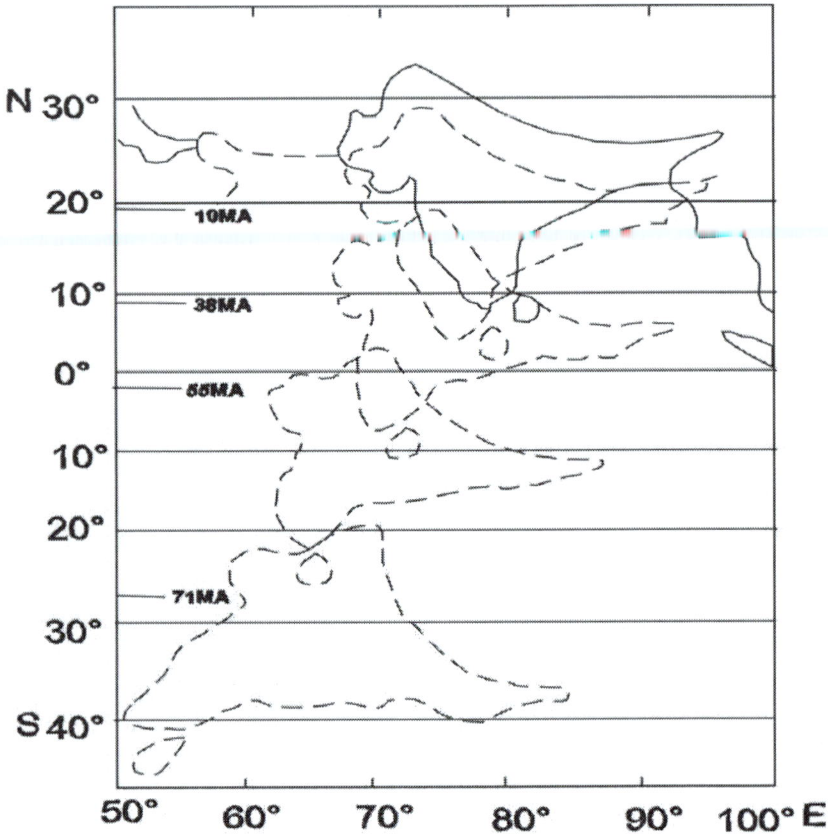

Fig. 14.3 Flight of Indian plate northwards

The collision of Indian Plate with Eurasian plate is visualised as a two-stage phenomenon as it envisages a continent to continent collision in a process where the intervening oceanic crust got consumed. The oceanic crust represented by Dras–Kohistan island arc first collided with the Indian Plate along Indus Suture Zone. Subsequently both of them moved northwards and collided with Eurasia along the Shyok Suture at a much later date. Simultaneously there was suturing along the eastern flank of the Indian Plate with the Mayanmar sub-plate.

14.4.2 Subduction and Obduction

One of the important lines of evidence supporting the subduction of Neo-Tethyan oceanic crust beneath the Dras island arc is the pre-collision high pressure and low temperature blueschist facies metamorphism in the Indus Suture Zone. The other line of evidence is the obduction of ophiolite blocks and ophiolite mélange on the Zanskar shelf before the closure of Tethyan marine sedimentation. Throughout the Himalayan region, obducted ophiolite complexes occur in two tectonic settings, i.e. in thrust sheets obducted southwards and disrupted lenses constituting the mélange. Spontang ophiolite klippe resting on Lamayuru Formation in Ladakh constitutes the example of the former while the belt of ophiolite mélange along Indus Suture Zone is the example of the latter.

After the closure of the ocean, a south-west directed thrusting followed, resulting in piggy-back riding of the Eurasian plate on the subducted northern margin of the Indian plate and the consequent crustal thickening. There is strong geophysical evidence supporting the crustal thickening in the Lhasa Block due to the subduction of the Indian plate underneath it.

14.5 RISE OF THE HIMALAYA

The collision related thrust structures were the precursors during the various phases of Himalayan orogeny that succeeded it. The first phase of this orogeny was marked by folding, uplift and thrusting due to the subduction. Eventually these resulted in the generation of main thrusts at various places, notably the set of thrusts constituting the Indus–Tsangpo Suture Zone (ITSZ) or what in Pakistan part is referred to as the Main Mantle Thrust (MMT), the Main Central Thrust (MCT) and the Main Boundary Thrust (MBT). These structures evolved around Oligocene when marine sedimentation was finally replaced by fresh water deposits (Murree and equivalents) in the frontal part of rising Himalaya (see Sec. 12.3.2). The second phase of orogeny resulted in the uplift of these rocks in Early Miocene sometime around 20 Ma. This phase was attended upon by intrusion of acid plutons and metamorphism of Central Crystallines. The Siwalik molasse basin evolved following this phase of orogeny and Lower and Middle Siwalik rocks were deposited. The third phase of orogeny was extremely profound and it resulted not only in the emergence of Central Crystalline axis as a main watershed of Himalaya but also resulted in the uplift of Pir Panjal Range generating the Karewa basin to the north. This phase of uplift took place towards the beginning of Late Pliocene and would be dated around 3.5 Ma. The last phase of Himalayan orogeny resulted in the Himalaya attaining the present stature and caused the uplift of Siwalik rocks and their folding. It also resulted in the formation of the Indo-Gangetic basin that is the present foredeep receiving sediments from Himalaya through the agency of various river systems. This phase took place in Pleistocene and could be dated between 0.8 and 0.5 Ma. The tectonic events following this phase resulted in the development of terraces and their tilting and thrusting of Siwalik rocks over the Indo Gangetic alluvium along what is known as Himalayan Frontal Thrust (HFT) (See Figs. 14.3 and 14.4). These are, however, within the fold of neotectonic activity that is continuing up to present.

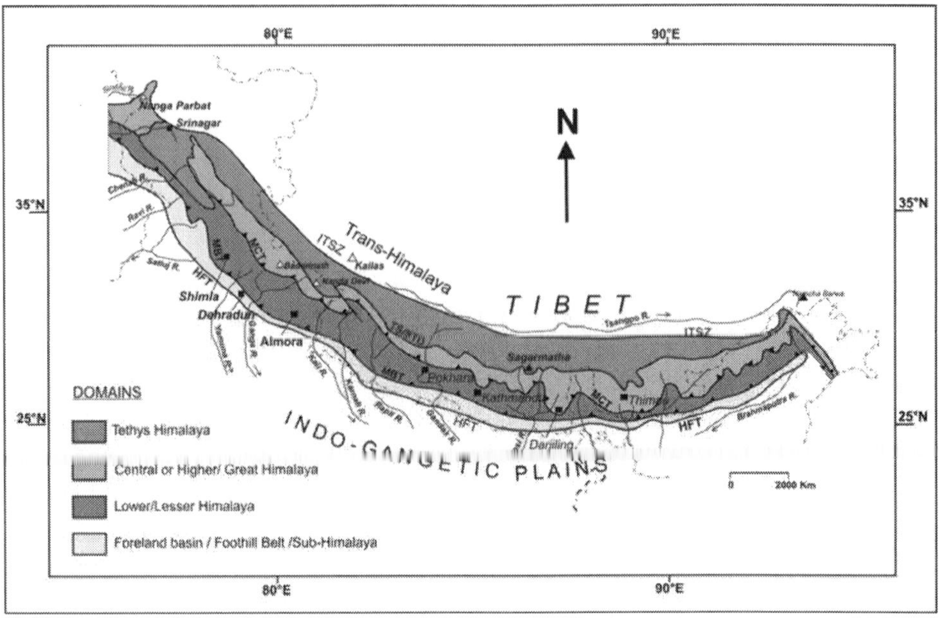

Fig. 14.4 Tectonic map of Himalaya

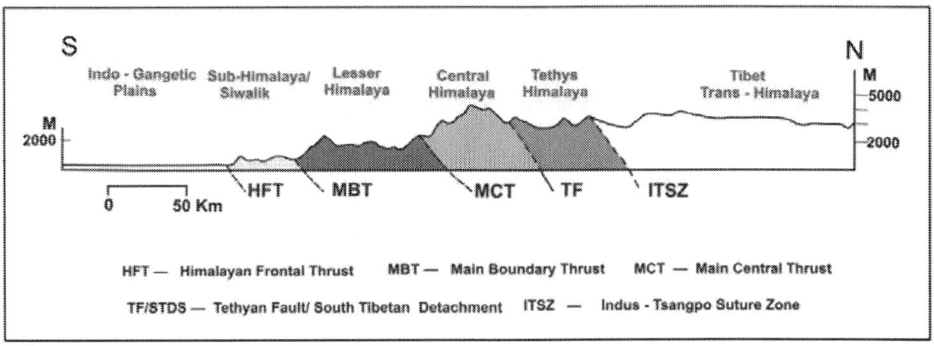

Fig. 14.5 Tectonic framework of Himalaya across a regional cross section

14.6 NEOTECTONIC ACTIVITY

Neotectonic activity has universally affected the entire Indian sub-continent to a greater or lesser degree. There are widespread evidences of the tectonic activity during Late Pleistocene and Holocene. They are reflected in morphological evolution of rivers and changes in the drainage patterns. Structural evidences include tilting of river terraces and angular unconformity between various levels of terrace deposits and alluvium.

In the Himalayan region, many rivers cutting across mountain barriers through gorges and canyons indicate a rapid rate of uplift. There are also evidences of Siwalik rocks overriding the Indo-Gangetic alluvium at several places. Monitoring of some of the Himalayan thrusts and faults for seismological purposes has indicated that they have been active in recent times, and the activity is continuing. Some of the more active thrusts include the Srinagar and Manora Thrusts of Uttarakhand, Riasi Thrust of Jammu and Soan Thrust in Subansiri river of Arunachal among many others.

In the coastal regions, the raised beaches are indicative of uplift of the order of several metres. The western Ghat escarpment is believed to be a recent fault line scarp. In the Saurashtra coast, the Pleistocene meliolite bearing rocks are uplifted along an active fault to a height of over 75 metres. Evidences of neotectonic activity in Narmada valley are widespread where Pleistocene sediments have been displaced along rejuvenated active faults. Many evidences have been recorded in other peninsular river basins.

One of the major changes that may be due to neotectonic activity was the disruption of the drainage pattern in parts of northern India. A major river, the mythical Saraswati, that was flowing through Haryana and Rajasthan into Rann of Kutch, and was joined by Sutlej and probably Yamuna, vanished and these tributaries fed Indus river system and Ganga river system, respectively, rendering large areas in Rajasthan and Kutch as desert regions. This could have been a combined effect of neotectonics and climate change.

Appendix - I

Phanerozoic Stratigraphic Time Scale

[Note: The gaps in the column of stage indicate that the stages have not been formally named. Global Stratotype Sections and Points (GSSPs) are shown only where they have been formally approved by the International Commission of Stratigraphy.]

CENOZOIC ERA/ERATHEM

PERIOD/ SYSTEM	EPOCH/ SERIES	STAGE/AGE	AGE (Ma)	GSSPs
Quaternary	Holocene			
	Pleistocene		1.806	Vrica, Calabria
Neogene	Pliocene	Gelasian	2.588	Monte San, Nicola, Sicily, Italy
		Placenzian	3.600	Pinta Picola, Sicily, Italy
		Zanclean	5.332	Eraclea Minoa, Sicily, Italy
	Miocene	Messinian	7.246	Oued Akrech, Rabbat, Morocco
		Tortonian	11.608	Monte dei Corvi Beach, Ancona, Italy
		Serravallian	13.65	
		Langhian	15.97	
		Burdigalian	20.43	
		Aquitanian	23.03	Lemme-Carrosio, N. Italy
Paleogene	Oligocene	Chattian	28.4	
		Rupelian	33.9	Massignano, Ancona, Italy
	Eocene	Priabonian	37.2	
		Bartonian	40.4	
		Lutetian	48.6	
		Ypresian	55.8	Dababiya, Luxor, Egypt
	Paleocene	Thanetian	58.7	
		Selandian	61.7	
		Danian	65.5	El Kef, Tunisia

MESOZOIC ERA/ERATHEM

PERIOD/ SYSTEM	EPOCH/ SERIES	STAGE/ AGE	AGE (Ma)	GSSPs
Cretaceous	Upper	Maastrichian	70.6	Tercis-les-Bains, Landes, SW France
		Campanian	83.5	
		Santonian	85.8	
		Coniacian	89.3	
		Turonian	93.5	Rock Canyon, Pueblo, Colorado, USA
		Cenomanian	99.6	Mont Risou, Rosans, Haute-Alpes, France
	Lower	Albian	112.0	
		Aptian	125.0	
		Barremian	130.0	
		Hauterivian	136.4	
		Valanginian	140.2	
		Berriasian	145.5	
Jurassic	Upper	Tithonian	150.8	
		Kimmeridgian	155.7	
		Oxfordian	161.2	
	Middle	Callovian	164.7	
		Bathonian	167.7	
		Bajocian	171.6	Cabo Mondego, W. Portugal
		Aalenian	175'6	Fuentelsaiz, Spain
	Lower	Toarcian	183.0	Robin Hood's Bay, Yorkshire, UK
		Pliensbachian	189.6	East Quantox Head, West Somerset, GB
		Sinemurian	196.5	
		Hettangian	199.6	
Triassic	Upper	Rhaetian	203.6	
		Norian	216.5	
		Carnian	228.0	
	Middle	Ladinian	237.0	
		Anisian	245.0	
	Lower	Olenekian	249.7	
		Induan	251.0	Meishan, Zhejiang, China

PALAEOZOIC ERA / ERATHEM

PERIOD / SYSTEM	EPOCH / SERIES		STAGE / AGE	AGE (Ma)	GSSPs
Permian	Lopingian		Changhsingian	253.8	
			Wuchiapingian	260.4	Penglaitan, Guangxi Province, China
	Guadalupian		Capitanian	265.8	Nipple Hill, Guadalupe Mountains, TX, USA
			Wordian	268.0	Guadalupe Mountains, TX, USA
			Rodian	270.6	Guadalupe Mountains, TX, USA
	Cisuralian		Kungurian	275.6	
			Artinskian	284.4	
			Sakmarian	294.6	
			Asselian	299.0	Aidaralash, Ural Mountains, Kazakhstan
Carboniferous	Penn-sylvanian	Upper	Gzhelian	303.9	
			Kasimovian	306.5	
		Middle	Moscovian	311.7	
		Lower	Bashkirian	318.1	Arrow Canyon, Nevada, USA
	Missis-sippian	Upper	Serpukhovian	326.4	
		Middle	Visean	345.3	
		Lower	Tournaisian	359.2	La Serre, Montagne Noir, France
Devonian	Upper		Famennian	274.5	Coumiac, Montagne Noir, France
			Frasnian	285.3	Col du Puech, Montagne Noir, France
	Middle		Givetian	391.8	Jebel Mech Irdane, Tafilalt, Morocco
			Eifelian	397.5	Wetteldorf Richtschnitt, Eifel Hills, Germany
	Lower		Emsian	407.0	Zinzil'ban Gorge, Uzbekistan
			Pragian	411.2	Velka Chuchle, SW Prague, Czeck Rep.
			Lochkovian	416.0	Klonk, SW of Prague, Czeck Rep.
Silurian	Pridoli			418.7	Pozary, Prague, Czeck Rep.
	Ludlow		Ludfordian	421.3	Sunnyhill, Ludlow, UK
			Gorstian	422.9	Pitch Coppice, Ludlow, UK

	Wenlock	Homerian	426.2	Whitewell Coppice, Homer, UK
		Sheinwoodian	428.2	Hughley Brook, Apedale, UK
	Llandovery	Telychian	436.0	Cefn Cerig, Llandovery, UK
		Aeronian	439.0	Trefawr, Llandovery, UK
		Rhuddanian	443.7	Dob's Linn, Moffat, UK
	Upper	Himantian	445.6	
			455.8	
			460.9	Fagelsang Scania, Sweden
Ordovician	Middle	Darriwillian	468.1	Huangnitang, Zhejiang Province, China
			471.8	
			478.6	Diasbasbrottet, Hunneberg, Sweden
	Lower	Tremadocian	488.3	Green Point, Newfoundland, Canada
Cambrian	Furongian			
		Paibian	501	Paibi, Hunan Province, China
	Series 3		513.0	
	Series 2		521.0	
	Terreneuvian		542.0	Fortune Head, Newfoundland, Canada

■■■■

Appendix - II
SELECTED BIBLIOGRAPHY

General

Balasubrahmanyan MN 2006 *Geology and Tectonics of India*, Memoir: International Association for Gondwana Research

Bhargava ON and Bassi UK 1999 *Geology of Spiti-Kinnaur, Himachal Himalaya*, Memoir: Geological Socity of India

Gaetani M, Casnedi R, Fois, E, Garzanti E, Jadoal F, Nicora A and Tintori A 1986 Stratigraphy of the Tethys Himalaya in Zanskar, Ladakh, *Rivista Italiana di Paleontologia e Stratigrafia* **91** 443-478.

Gansser A 1964 *Geology of Himalayas* Interscience Wiley

GSI (Geological Survey of India) 1971 Code of Stratigraphic Nomenclature of India: Geological Survey of India Miscellaneous Publication **20**

Gopender Kumar 2005 *Geology of Uttar Pradesh and Uttaranchal,* Geological Society of India

Krishnan MS 1982 *Geology of India and Burma,* 6th Edition CBS Publishers

Merh SS 1995 *Geology of Gujarat* Geological Society of India

Pascoe EH 1959 & 1964 *A Manual of Geology of India and Burma, (Vol.1, 2 & 3),* 3rd Edition Govt. of India Press

Srikantia SV and Bhargava ON 1998 *Geology of Himachal Pradesh,* Geological Society of India

Thakur VC 1993 Geology of Western Himalaya, Pergamon

Vaidyanadhan R and Ramakrishnan M 2008 *Geology of India (Vol. 1 and 2),* Geological Society of India

Valdiya KS 2010 *The Making of India – Geodynamic Evolution,* Macmillan Publishers (India)

Wadia DN 1975 *Geology of India,* 4th Edition Tata McGraw Hill

Chapter 1 Stratigraphic Concepts

Ager DV 1981 The nature of the stratigraphical record, 2nd Edition Macmillan

Boggs Sam 2008 Principles of Sedimentology and Stratigraphy, Pearson Education.

Cubitt JN and Reyment RA (eds) 1982 *Quantitative Stratigraphic Correlation*, John Wiley & Sons.

Donald R, Prothero and Schwab Fred 2004 Sedimentary Geology: An introduction to sedimentary rocks and stratigraphy, Freeman

Dunbar CO and Rodgers J 1957 *Principles of Stratigraphy,* John Wiley & Sons

Grastein WC, Ogg J and Smith A 2004 *A Geologic Time Scale,* Cambridge Univ. Press

Krumbein WC and Sloss LL 1963 *Stratigraphy and sedimentation*, 2nd Edition: Freeman

Lemon RR 1990 *Principles of Stratigraphy,* Merrill Publishing Company

Schoch RM 1989 *Stratigraphy, Principles and Methods,* New York: Van Nostrand Reinhold:

Chapter 2 Physiography of India

Gopal Singh 1976 *A Geography of India*, Atma Ram and sons

Chapter 3 Peninsular Archean and Proterozoic Crystalline Rocks

Gupta SN, Arora YK, Mathur RK, Iqballuddin, Prasad B, Sahai TN and Sharma SB 1997 *The Precambrian Geology of the Aravalli region, southern Rajasthan and northeastern Gujarat*, Memoir: Geological Society of India

Naqvi SM and Rogers JJW *Precambrian Geology of India,* Oxford University Press

Paliwal BS (ed) 1998 *The Indian Precambrian,* Jodhpur: Scientific Publishers

Radhakrishna BP and Vaidyanadhan R 1997 *Geology of Karnataka,* 2nd Edition: Geological Society of India

Roy AB (ed) 1988 *Precambrian of Aravalli mountain, Rajasthan, India,* Memoir: Geological Society of India

Saha AK 1994 Crustal evolution of Singhbhum-North Orissa, Eastern India, Memoir: Geological Society of India

Sarkar G, Corfu F, Paul DK, McNaughton NJ, Gupta SN and Bishul PK 1993 Early Archaean crust in Bastar craton, central India: a geochemical and isotopic study, *Precambrian Research* **62** 127-137.

Swami Nath J and Ramakrishnan M (eds) 1981 *Early Precambrian Supracrustals of Southern Karnataka,* Memoir: Geological Society of India

Chapter 4 Proterozoic Peninsular sedimentary basins:

Auden JB 1933 *Vindhyan sedimentation in Son Valley,* Memoir: Geological Society of India

Bhattacharyya A (ed) 1996 *Recent advances in Vindhyan Geology,* Memoir: Geological Society of India

Chakraborty C 2006 Proterozoic intracontinental basin: the Vindhyan example, *Journal Earth System Sciences* **115** 3-22.

Chakraborty C, Sarkar S, Chaudhuri AK and Dasgupta S 1996 Depositional environment of Vindhyan and other Purana basins: A reappraisal in the light of recent findings, *Memoir Geological Societyof India* **36** 101-126.

Das N, Dutta DR and Das DP 2001 Proterozoic cover sediments of south-eastern Chhattisgarh State and adjoining parts of Orissa, *Geological Survey of India Special Publication* **55(2)** 237-262.

Jaya Prakash AV 2007 *Purana basins of Karnataka,* Memoir: Geological Survey of India

Kale VS and Phansalkar VG 1991 Purana basins of Peninsular India: A review, *Basin Research* **3** 1-36.

Kathal PK, Patel DR and Alexander PO 2000 An Ediacaran fossil *Spriggina* from the Semri group and its implication on the age of the Proterozoic Vindhyan Basin, Central India, *Neues Jahrbuch fur Geologie Paläontologie: Mannheim Germany* **6** 321-332.

Lakshminarayana G, Bhattacharjee S and Ramanaidu KV 2001 Sedimentation and stratigraphic framework of Cuddapah basin, *Geological. Survey of India Special Publication* **55(2)** 31-58.

Radhakrishna BP (ed) 1987 *Purana basins of Peninsular India,* Memoir: Geological Society of India

Singh IB 1985 Palaeogeography of the Vindhyan basin and its relationship with other Late Proterozoic basins of India, *Journal of Palaeontological Society of India* **30** 35-41.

Chapter 5 Precambrian Rocks of Himalaya

Shankar R, Kumar G, Mathur VK and Joshi A 1993 Stratigraphy of Blaini, Infra-Krol, Krol and Tal successions, Krol Belt, Lesser Himalaya, *Indian Journal of Petroleum Geology* **2** 99-136

Chapter 6 Phanerozoic sedimentation and transgressive events

Bhargava ON 2008 *Paleozoic Successions of the Indian Plate,* Memoir: Geological Society of India

Brookfield ME 1993 The Himalayan passive margin from Precambrian to Cretaceous times, *Sedimentary Geology* **84** 1-35

Klootwijk CT, Shah SK, Gergan J, Sharma ML, Tirkey B and Gupta BK 1983 A palaeomagnetic reconnaissance of Kashmir, northwestern Himalaya, India, *Earth and Planetary Science Letters* **63**.305-324

Myrow PM, Hughes NC, Paulsen TS, Williams IS, Parcha SK, Thompson KR, Bowring SA, Peng SC and Ahluwalia AD 2003 Integrated tectonostratigraphic analysis of the Himalaya and implications for its tectonic reconstruction, *Earth and Planetary Science Letters* **212** 433-441

Shah SK 1991 Stratigraphic setting of the Phanerozoic rocks along the northern boundary of the Indian plate, *Physics and Chemistry of the Earth* **18(2)** 317-328

Chapter 7 Ediacaran and Cambrian sedimentary cycle

Bhatt DK 1991 The Precambrian-Cambrian transition interval in Himalaya with special reference to small shelly fossils – a review of current status of work, *Journal of the Palaeontological Society of India* **36** 109-120

Garzanti E, Casnedi R and Jadoul F 1986 Sedimentary evidence of a Cambro-Ordovician orogenic event in the northwestern Himalaya, *Sedimentary Geology* **48** 237-265.

Hughes NC 2016 The Cambrian Palaeontological record of the Indian Subcontinent, *Earth-Science Reviews* **159** 428-461

Hughes NC, Peng S, Bhargava ON, Ahluwalia AD, Walia S, Myrow PM and Parcha SK 2005 Cambrian biostratigraphy of the Tal Group, Lesser Himalaya, India and early Tsanglangpuan (late early Cambrian) trilobites from the Nighali Dhar syncline, *Geological Magazine* **142(1)** 57-80.

Jell PA and Hughes NC 1997 Himalayan Cambrian Trilobites, *Palaeontology (Special paper)* **58**. 1-113

Myrow PM, Thompson K, Hughes NC, Paulsen TS, Sell BK and Parcha SK 2006 Cambrian stratigraphy and depositional history of Northern Indian Himalaya, Spiti Valley, north- Central India. *Bulletin Geological Society of America* **118(3/4)** 491-510.

Shah SK 1982 Cambrian Stratigraphy of Kashmir and its boundary problems, *Precambrian Research* **17** 87-98

Shah SK 1993 Cambrian biofacies and Faunal Provinces of Himalaya, *Journal of the Palaeontological Society of India* **38** 37-42.

Chapter 8 Post-Cambrian Paleozoic stratigraphy

Draganits E, Mawson R, Talent J and Krystyn L 2002 Lithostratigraphy, conodont biostratigraphy and depositional environment of the Middle Devonian (Givetian) to Early Carboniferous (Tournaisian) Lipak Formation in the Pin Valley (NW India), *Rivista Italiana di Paleontologia e Stratigrafia* **108(1)** 7-35.

Ranga Rao A, Dhar CL, Rao SV and Shah SK 1982 Contribution to the stratigraphy of Spiti, *Himalayan Geology* **12** 98-113.

Talent JA, Goel RK, Jain AK and Pickett JW 1988 Silurian and Devonian of India, Nepal and Bhutan; biostratigraphic and paleobiogeographic anomalies, *Courier Forschungs Institut Senckenberg* **106** 1-57.

Chapter 9 Gondwana Sedimentation

Casshyap SM 1970 Sedimentary Cycles and environment of deposition of the Barakar coal measures of Lower Gondwana, India, *Journal of Sedimentary Petrology* **40** 283-303

Casshyap SM and Tiwari RC 1984 Fluvial models of the Lower Permian Gondwana coal measures of Son-Mahanadi and Koel-Damodar basins, in *Sedimentology of coal and coal bearing strata* (eds.) R. A. Rahmani and R. M. Flores, Special Publication of the International Association of Sedimentologists **7**121-147 pages

Lele KM 1964 The Problem of Middle Gondwana in India, Proceedings. XXII International Geological Congress: Geological Survey of India **9** 181-200 pages

Mukhopadhyay G, Mukhopadhyay SK, Roychowdhury M and Paeui PK 2010 Stratigraphic Correlation between different Gondwana basins of India, *Journal of Geological Society of India* **76** 251-266.

Venkatachala BS and Maheshwari HK (eds) 1991 *Indian Gondwana*, Memoir: Geological Society of India

Chapter 10 Mesozoic marine sequence

Jai Krishna 1987 An overview of the Mesozoic stratigraphy of Kachchh and Jaisalmer basins, *Journal of Paleontological Society of India* **32** 136-152

Sahni Ashok (ed) 1996 *Cretaceous Stratigraphy and Palaeoenvironments,* Memoir: Geological Society of India

Chapter 11Deccan Volcanics and Related sediments

Subbarao KV (ed) 1999 *Deccan Volcanic Province,* Memoir: Geological Society of India

Tandon SK, Sood A, Andrews JE and Dennus PF 1995 Palaeoenvironments of the dinosaur bearing Lameta beds (Maastrichtian), Narmada valley, central India, *Palaeogeography. Palaeoclimatology. Palaeoecology* **117** 153-184.

Chapter 12 Cenozoic Sedimentation

Bhatt DK 1989 The lithostratigraphy of Karewa Group, Kashmir Valley, India and a critical review of the fossil record, Memoir: Geological Survey of India

Biswas SK and Raju DSN 1973 The rock stratigraphic classification of the Tertiary sediments of Kutch, *Bulletin Oil and Natural Gas Commission* **10(1-2)** 33-46

Das Gupta AB and Biswas AK 2000 *Geology of Assam*, Geological Society of India

Madan Mohan 1982 Palaeogene Stratigraphy of Western India, *Palaeontological Society of India Special Publication* **1** 21-36.

Narayana AC (ed) 2002 *Late Quaternary Geology of India and sea level changes,* Memoir: Geological Society of India

Sahni A and Vimal Kumar 1974 Palaeogene Palaeobiogeography of the Indian subcontinent, *Palaeogeography, Palaeoclimatology, Palaeoecology* **15(3)** 209-226

Sahni A and Mitra HC 1980 Neogene palaeogeography of the Indian sub-continent, *Palaeogeography, Palaeoclimatology, Palaeoecology* **31** 39-62

Chapter 13 The Siwalik and related Sedimentation

Nanda AC 2000 Siwalik Group: Palaeontologic and magnetostratigraphic aspects, *Geological Survey of India Miscellaneous Publication* **64** 53-58

Opdyke ND, Lindsay E, Johnson GD, Tahirkhelli RAK and Mirza MA 1979 Magnetic polarity stratigraphy and vertebrate palaeontology of the Upper Siwalik sub-group of Northern Pakistan, *Palaeogeography, Palaeoclimatology, Palaeoecology* **27** 1-34

Ranga Rao A, Agarwal RP, Sharma UN, Bhalla MS and Nanda AC 1988 Magnetic polarity stratigraphy and vertebrate palaeontology of the Upper Siwalik Subgroup of Jammu Hills, India, *Journal of Geological Society of India* **31(4)** 361-385

Tandon SK 1991 The Himalayan Foreland: Focus on Siwalik basin, in *Sedimentary Basins in India* (eds.) SK. Tandon, CC Pant and SM Casshyap, Nainital: Gyanodaya Prakashan, 171-201 pages

Chapter 14 Tectonic evolution of Indian sub-continent

Klootwijk CT, Sharma ML, Gergan J, Shah SK and Gupta BK 1986 Rotational over thrusting of the northwestern Himalaya: further palaeomagnetic evidence from the Riasi thrust sheet, Jammu foothills, India, *Earth and Planetary Science Letters* **80** 375-393

Owen HG 1983 Atlas of continental displacement, 200 million years to the present, Cambridge University Press.

Radhakrishna BP and Merh SS 1997 *Vedic Saraswati — Evolutionary history of a lost river of NW India,* Memoir : Geological Society of India.